No waste like home

THE PLAN THAT WILL CHANGE YOUR WORLD

Penney Poyzer

First published in Great Britain in 2005 by
Virgin Books Ltd
Thames Wharf Studios
Rainville Road
London
W6 9HA

A catalogue record for this book is available from the British Library.

ISBN 0 7535 1027 8

The paper used in this book is a natural, recyclable product made from wood grown in sustainable forests. The manufacturing process conforms to the regulations of the country of origin.

Designed by Smith & Gilmour, London

Printed and bound in Great Britain by Butler and Tanner Ltd.

CONTENTS

Introduction

The filming of **No Waste like Home** and the production of this book has been a wonderful personal journey for me. At 45 you don't expect to embark on a new career, but it happened to me in July 2004 when I got a phone call to see if I would be interested in becoming a presenter. Opportunities like that don't hit you on the head very often, so with the encouragement of my husband Gil, mum and friends, I went off on location for six months, with occasional visits home. It was an amazing life-changing experience.

It has taken me 20 years or so to work out how to balance my desire for a full and interesting lifestyle with my passion to look after our beautiful planet, but it is possible. Hopefully, this book will help you to achieve that balance, too.

The main message I want to get across is that we need to get wise to waste and start changing our wasteful ways.

In order to do this we need to change the way we think, shop, consume and live. But that doesn't mean depriving ourselves of life's little luxuries – we'd be mad to do that, right? Well, of course we would.

It's time for us all to face facts – we're in the middle of a waste crisis and we're to blame. We're creating more waste than ever before and this mountain of muck is increasing faster than recycling can keep up.

Living in a wasteful way harms your pocket and the environment. If you're paying for more things than you need to, only to chuck them into the rubbish bin at the end of the week, then you need to wake up to waste.

I'm not about to tell you that you've got to give up your cars or become vegetarians. I simply want to convince you that being less wasteful just means changing your habits and thinking more about what and why you do things.

I've made it my life's ambition to get people to stop and think. If we just took stock for a minute to think about why we did certain things, and to ask why we let ourselves be seduced into buying extra goodies at the supermarket checkout, then we could start to make positive changes to our lives. I want you to start living for tomorrow instead of spending for today.

We've become slaves to the advertisers and manufacturers – buying into the promise of a better, cleaner, leaner lifestyle if we purchase this or that gadget; or if we owned one of these bright shiny gizmos that guzzle up all our hard-earned cash in energy bills. If we don't change our ways and start living less wasteful lives then, frankly, we're done for. This isn't just about you; it's about your kids, your grandchildren, and your grandchildren's children. It's about safeguarding their futures too.

You only have to look back at the generation before last to see how wasteful we've become over the years. If we were doing what granny did, then we'd automatically improve our lives as well as the environment. Why? Because granny, as a matter of course, didn't bat an eyelid at having to tip all the vegetable peelings on to the compost heap, or washing nappies by hand. It was normal practice for granny to head for

the local shops armed with returnable glass bottles and come back home with a small deposit in return. Granny would think nothing of taking the iron to the local electrical shop to get it mended, rather than chucking it in the rubbish bin – we could all learn a thing or two from granny.

The average family wastes £430 on food, and generates around one tonne of household rubbish and waste per year. They can also spend as much as £301.60 a year on electricity, £265.20 on gas and £280 on water, but, by making a few simple changes, you can cut you and your family's bills by at least 50%.

A staggering £200 is spent per household, per year through needlessly wasting energy. In the UK, nearly 30% of all energy supplied is used in our homes, so it's important that we don't waste it. After all, our fossil fuels are not going to last for ever.

I want to instill a sense of passion, urgency and awareness in you so that you can get active. By making positive life changes, you can reap the rewards of more wealth and better health, lead fuller, happier lives by spending time with your friends and family, and worry less about the pennies by making the right choices. Being unhappy, stressed and unhealthy reduces your ability to get the most out of life and jeopardises your chances of doing the things that you find genuinely rewarding and fulfilling.

I guarantee that if you follow the tips you'll find throughout this book, you'll radically improve the quality of your life and all those around you. The majority of things in our rubbish can be recycled and composted and this would reduce the amount of waste that is buried or burned. This, in turn, would reduce pollution and community charges.

This isn't just about you;
it's about your kids,
your grandchildren,
and your grandchildren's children.
It's about safeguarding their futures too.

All you need to do to join the rubbish revolution is apply some simple Lifestyle Rules:

1 Reduce your rubbish
Using less means we throw away less

2 Reuse bags, bottles, jars, paper
and other materials rather than tossing them in the bin

3 Recycle glass, cans, metal,
textiles, batteries, oil, furniture and electrical goods

What do you need to get you started? Well, you've got the first bit of kit, which is this book. You'll not only find lots of no-cost, low-cost tips on how to reduce waste in each of the chapters, but you'll find a whole host of other sources of information telling you where you can buy certain products or where you can find out more about eliminating waste.

The best news of all is that there's no need to lay out vast amounts of cash to get you started. The only thing you need to start making these positive life changes today is YOUR BRAIN!

CHAPTER ONE

No waste like home

Have you ever stopped to think about how much rubbish you and your family throw away every week?

Or why we need to stop throwing so much away? Do you even know where your 'waste' goes to or what technically counts as 'waste'?

Well, here are some facts to get you thinking . . .

HOUSEHOLD HOWLERS

GENERAL GARBAGE

- The UK produces more than 434 million tonnes of waste every year; at that rate we could fill the Albert Hall in London with rubbish in less than two hours.
- Every year UK households throw away the equivalent of three and a half million double-decker buses (almost 30 million tonnes), a queue of which would stretch from London to Sydney (Australia) and back.
- The UK creates 9 million tonnes of packaging debris a year which is enough to fill 13,000 jumbo jets.
- Around £430 worth of food a year per person is thrown away, with the UK population currently around 59.9 million.
- On average, each person in the UK throws away seven times their own body weight in rubbish every year.

PLASTIC – NOT SO FANTASTIC!

- Every year, an estimated 17 billion plastic carrier bags are given away by supermarkets; this is equivalent to over 290 bags for every person in the UK.
- We produce and use 20 times more plastic today than we did 50 years ago.

REVOLTING RUBBISH

- Babies' disposable nappies make up about 2–4% of the average household rubbish bin. This equates to 400,000 tonnes sent to landfill each year.
- Dealing with nappy waste costs individual local authorities hundreds of thousands of pounds a year.
- In the UK over eight million disposable nappies are thrown away each day.

OODLES OF OIL

- One litre of oil can pollute one million litres of fresh drinking water.
- Waste oil from nearly three million car oil changes in Britain is not collected; if collected properly, this could meet the annual energy needs of 1.5 million people.

PREPOSTEROUS PAPER

- About one fifth of the contents of household dustbins consists of paper and card, of which half is newspapers and magazines – this is equivalent to over 4 kg of waste paper per household in the UK each week that could have been recycled.

GLORIOUS GLASS

- On average, every family in the UK consumes around 330 glass bottles and jars a year. It is not known how long glass takes to break down, but it is so long that glass made in the Middle East over 3,000 years ago can still be found today.

- Recycling two bottles saves enough energy to boil water for five cups of tea.

Main Source: www.wasteonline.org.uk
Other Sources: British Gas, Scottish Oil Care Campaign, Women's Environmental Network, *Which?*, Environment Agency Nappy Life Cycle Analysis Report, (May 2005)

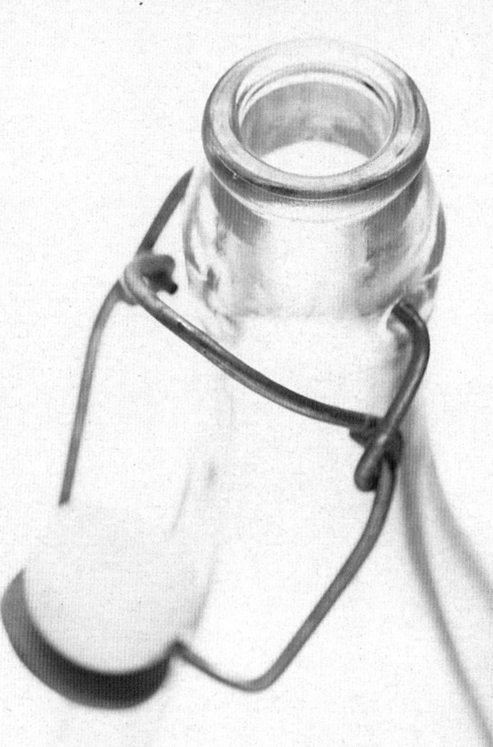

DOWN IN THE DUMPS

So, what's stopping us?

Everywhere we look there's waste, and we're the ones who create it. There's no such thing as the garbage fairy who mysteriously comes along to fill up our bins with rubbish when we're not looking – it's us.

Our lives are becoming increasingly busy and time-poor and we're cramming more and more into hectic schedules. As a result, we're becoming ever more reliant on the cult of convenience. But, the trouble with convenience is that it often goes hand in hand with waste.

We need to rethink our approach if we're to conserve natural resources, maintain our health and minimise waste disposal charges on local communities. Yes, that's right. Getting rid of our rubbish costs money and we're the ones who foot the bill so it's in all of our best interests to do something about it.

At the start of the twentieth century bigger factories started churning out a whole range of products designed to make our lives easier. Over the years more of us have bought in to this 'luxury' lifestyle, filling our homes with TVs, computers, mobile phones, games consoles, washing machines, super-duper spin-dryers and whizzy food mixers – the list is endless.

Because we're so obsessed with having 'this year's model' all these 'essential' household appliances are replaced at a truly alarming rate with brand-spanking-new ones. This, in turn, means that millions of tonnes of yesterday's technology end up as waste.

What's more, we're poisoning the air around us through the harmful waste gases belching from power stations due to our wasteful and ever more demanding home energy consumption. We forget that we're not just emitting poison into the atmosphere; we're creating pollution in the smaller context of our homes too. Thousands of household fumes and toxins are right under our noses creating a great big stink in our own personal space. We pollute the air with toxic car-exhaust fumes also. Add to that the sewage sludge and other liquid waste that stream out of our pipes and it's not hard to see that we're slowly being engulfed by our own waste.

If we minimise our household waste, we could cut our household bills by half.

THINGS YOU SHOULD KNOW ABOUT THE PLANET

OR WHAT I CALL 'THE GREEN BIT'

Over the past decade abnormal weather conditions have made headline news in many parts of the world. For centuries Britain had mild summers and winters and moderate rainfall and winds. But now our weather is becoming increasingly unpredictable and seven out of ten of the warmest years on record have occurred in the last decade.

In the past, climate change has been driven by natural events and processes, but in more recent times – since the industrial revolution – we have been putting more harmful gases into the Earth's atmosphere as a result of burning fossil fuels. Scientists now believe that, as a result, the atmosphere is trapping too much heat and the Earth is warming up. That may sound like good news if you're sick of the chilly old winter weather we get here in the UK – but it's not.

The lifestyle we have today is possible because we use energy from fossil fuels such as coal, oil and gas which are the main fuels used to generate electricity. And in our homes we use this energy to keep us warm and light our homes; cook and chill our food; and power a whole host of other appliances which make our lives 'easier' and more convenient. The only problem is that as we demand more and more fuel, we are pumping more pollution into the atmosphere causing problems such as global warming and acid rain.

WHAT IF WE DON'T CHANGE OUR WAYS?

Nobody's really sure of the full impact global warming will have on our lives, but there is growing evidence from scientists that it is a very real and serious problem. Experts predict that we will see more extreme and unpredictable weather conditions – including more flooding.

Heavy downpours in the UK now contribute twice as much rain as they did in the early 1960s. Increased air pollution will result in more breathing problems such as asthma, and changes in soil quality and water shortages will affect our ability to grow particular types of crops – which will have a knock-on effect on the type and price of food we buy and eat. Some predictions indicate that several coastal towns could be washed away by the effects of global warming.

United Nations scientists further predict that higher temperatures in Britain over the next 50 years may cause 10,000 more cases of food poisoning; 3,000 extra deaths annually from heatstroke; an extra 5,000 deaths from skin cancer; and 2,000 more cataract operations each year.

Taking small actions can really make a difference to the planet. If you just take on board one or two things from each chapter and practice them on a regular basis, you really will be making a big difference.

MY THREE LIFESTYLE RULES

Your first step is to follow three simple lifestyle rules. 80% of all household waste can be reduced, reused or recycled. This is your new mantra for a less wasteful lifestyle.

1 Reduce

- Aim to cut the amount of waste we produce.
- Donate old magazines to doctors' and dentists' waiting rooms.
- Buy 'loose' products to cut down on packaging.
- Buy concentrated products which use less packaging.

2 Reuse

- Further cut down the waste we produce by reusing certain things.
- Rent or borrow items we don't use very often, e.g. party decorations and crockery – some supermarkets hire out glasses for parties, saving on disposable cups.
- Donate old computer and audio-visual equipment to community groups or schools.
- Buy rechargeable items instead of disposable ones, e.g. batteries and cameras.
- Buy things in refillable containers, e.g. washing powders.

3 Recycle

- Take old clothes and books to charity shops, or have a car-boot sale.
- Look for long-lasting (and energy-efficient) appliances when buying new electrical items – ensure these are well maintained to increase product life cycle.
- If your local authority collects recyclable materials (e.g. paper, glass) as part of its waste collection service, then take advantage!
- Some wrapping paper can be recycled – but only if it is plain white with simple printing.
- Make a visit to a bottle, can or other type of recycling bank part of your routine – avoid making a special trip if possible.
- When you're recycling cans, squash them so they take up less room – and wash out food cans first.
- Even damaged clothes can be turned into cloths, furniture padding, or new fabric, so recycle these at clothes banks or via charity bags at your doorstep.
- Don't dump old furniture – look up details of furniture recyclers in your local phone book.
- Put the right materials in the right recycling bin or recycling bank; remember – check before you chuck!

CASE STUDY

BENDALL JONES FAMILY
Gemma, Mark and their daughters Verity 22 months, Poppy and Tess (identical twins) 8 months

When I started working with the participants in BBC2's No Waste Like Home TV series, one of the most common complaints I'd hear before they started on the road to a less wasteful life was that 'all this green stuff costs money'. Well, that's just plain wrong!

In fact what struck me most about all of the families I worked with on the show was how much money they were literally throwing away. Many families were stressed, tired and not communicating with each other because they were always stuck in front of the TV or games consoles. Their eating habits were atrocious because they relied on a diet of ready-meals and takeaways.

Take the Bendall Jones family, for example. Despite having their hands full with three nappy-filling children, they really could not see how the waste from their home was affecting the environment. Mark and Gemma were a typical young couple who believed, like most of us, that the luxuries they saw as an essential part of their busy lifestyle, were no big deal. They used 8–10 bags of rubbish per week and had never even heard of landfill.

Gemma didn't think there was a problem with the environment because she never heard about it on the news. By simply wanting to emulate the 'warm lighting' she saw in home improvement TV programmes, she was tripling her electricity bill. She also went out and left the TV on so that there would be a welcoming noise for the children when they got home. The washing machine was used for two lots of two-hour washes a day and she used the tumble dryer even though they had space for a washing line in the garden.

Mark was a handy man around the home but the house was not insulated properly so they had very large gas bills. And while Mark tried to keep the room thermostat turned down, Gemma feels the cold so they were constantly battling behind each other's backs.

However, the best thing about helping out with this family was the thought of stopping those nappy mountains. They use around 100 nappies a week. Gemma lobbed them around the house and Mark then went around and collected them when he got back from work. They called this 'poo patrol'. By realising their wasteful ways they were not only helping to improve the planet for their lifetime; they were making it a better place for their three beautiful children to grow up in.

Faced with the shocking truth I set about giving them a complete lifestyle makeover and watched them morph from wasteful miseries into happy, healthy and wealthy beings!

WASTE WEIGH-IN

I confronted each family from the show with an array of bin bags of their rubbish and, handing them a pair of rubber gloves each, informed them that they were going to sort through every single one.

We separated every item and put them into containers labelled 'reduce', 'reuse' and 'recycle', and gave them an on-the-spot tutorial on how to manage their household rubbish.

They were less than enthusiastic at first, but by the time I had finished with them, they were recycling converts, especially when they saw they could save money in the process!

ENERGY DRAIN

We then looked at how much energy the family was wasting, including through the use of their cars, electricity, water and heating.

TOXIC TESTS

I revealed the toxic cocktail of cleaning products under the sink in their kitchens and showed them how to use simple ingredients such as vinegar, newspaper and lemon juice to keep their homes clean and sparkling.

What struck me about all of the families I worked with on the show was how much money they were literally throwing away.

WHAT KIND OF WASTER ARE YOU?

There are many types of waster and we're all guilty of wasting to some extent. Tick any description from the following list that applies to you and see what your potential saving could be.

TOSSERS

Tossers throw out endless bins of rubbish without a second glance. You never apply the '3 Rs', so you've never thought about what could be reduced, reused or recycled.

GUZZLERS

Guzzlers are the big energy wasters. You leave lights blazing away in unoccupied rooms, you've got a glut of kitchen and other household electrical gadgets going all day long, and you leave the TV and video on constant stand-by.

BELCHERS

Belchers are the car users and major polluters. You may even have more than one car and you'd never, ever walk anywhere let alone think about what damage your fuel is causing.

PLONKERS

Plonkers are large families producing massive overflowing nappy mountains, munching their way through packets and packets of processed food, and binning large quantities of plastic and polystyrene take-away packaging.

SCRAPPERS

Hey big spenders! Who can't resist buying the latest trendy new clothes, make-up or shoes, then? I bet you never wear or use half the stuff you buy and once the latest TV 'fashion guru' tells you it's no longer hip, it'll go straight to the tip.

0–1

You're a pretty cool, green-thinking dude. You care about the planet in the right way and you try to do your bit. You probably recycle your newspapers and buy organic produce from the supermarket. You may well be ready for the next step – which means a potential for further savings. In fact, you notch up a POTENTIAL SAVING of £1,000.

2–3

You're sort of on the right track, but you tend to make excuses. It takes too much time to always recycle your magazines, so occasionally you'll throw them in the rubbish bin. And you just can't face not using the car – not even for a day. You may be shocked to discover that your POTENTIAL SAVING could be as much as £2,500 a year.

4–5

You really are the pits! You're hot water hogs; energy bandits and through and through wonton wasters. Well, sit up and pay attention. By changing your horrible habits you could make a POTENTIAL SAVING of £4,000 a year. Just think what you could spend it on instead of chucking it away – which is literally what you're doing by being so wasteful.

TOP TIPS WATCHING WASTE AT HOME

Talk about ways of **reducing your household waste** with your family and friends

Where you can, **try to mend or repair things,** rather than toss them away – if they have to go, at least **see if they can be recycled**

Read the labels on household detergents, cleaners and other chemicals and try to go for environmentally friendly alternatives

Use paper where possible (rather than plastic) as it can be recycled more easily

Around the house make sure you **have proper insulation,** turn off unnecessary lights, use low-energy and long-life bulbs, and buy energy-efficient machines and appliances

CHAPTER TWO

The rudiments of rubbish

We know that we have to sort our household rubbish into what can be recycled, but how do we know what to recycle and how to recycle it?

Buying 'greener' products which can be reused or recycled will help to reduce the rubbish that we produce, but sometimes buying 'green' products can be confusing.

Do you know what the symbols mean?

And how can you tell if one product contains more recycled material than another?

THE WASTE ROUTE

Tipping the balance

We already know that every household creates far too much waste, but what happens to it once it's collected by the binmen? Well, it's usually joined by waste collected from parks, schools, factories, businesses and shopping centres. All this waste is known collectively as 'municipal waste'.

In general, all municipal waste gets sent to landfill sites, which are essentially big holes in the ground that our rubbish gets tipped into and buried – that is until they can't squeeze any more in.

There are now thousands of managed landfill sites in the UK, but there are problems – and it's not just that space for landfill is running out. Landfill sites are smelly and even though our rubbish is buried underground, it produces large amounts of toxic gases that are released into the atmosphere. And then there's the excess rubbish that's carried off in the wind and attaches itself to the surrounding countryside. Landfill sites are also a haven for vermin such as rats, which spread disease. If all this isn't bad enough, liquid can seep into the soil from older sites – liquid that contains harmful chemicals such as pesticides, solvents and heavy metals, which contaminate our water supplies.

Some waste from sewage sludge is also buried in landfill sites, along with waste from mining and quarrying. As landfill waste decomposes, methane gas is released – it's currently estimated that over 1.5 million tonnes of methane are released by landfill sites in the UK each year. Methane contributes to global warming, which means we're likely to see more erratic weather conditions, more pollution and more of us suffering from respiratory problems such as asthma. However, methane is mostly managed in landfill sites and used to produce energy. Three quarters of the world's methane actually comes from animals such as cows, sheep and pigs passing wind.

Dust from landfill sites has also been known to cause skin and eye irritation to those living nearby.

Many areas of the country now have less than 10 years of landfill space left and the quantity of rubbish we produce is rising by at least 3% a year nationally.

England disposes of the majority of household waste through landfill and incineration with only 17% for recycling. The cost of managing household waste is forecast to double from £1.6 billion to £3.2 billion a year by 2020. All of us have a major part to play in helping to reduce our rubbish and increase recycling and composting. It is essential to safeguarding our long-term health and wellbeing.

A BURNING ISSUE

Some waste is burned in specially built incinerators; of the 7,000 incinerators in England and Wales, 12 burn municipal waste. However, the ashes that are left over from this process are then buried in a landfill site!

Incinerators produce smoke, gas and fumes that contain harmful chemicals, although the jury is still out on what these harmful chemicals will do to the environment in the future.

TOXIC TIPPING

Toxic waste harms the environment even in very small quantities, because many toxins don't break down once they're buried. So, there's always the danger that they will get passed back to us through the food that we buy, the air that we breathe and in the water that we drink.

THE BOTTOM LINE

Almost a third of the water we use at home is for flushing human waste (and chemicals we use to keep our loos clean) down our toilets. This waste can easily amount to 100 litres of water per person per day. This waste usually flows along pipes to the sewage treatment works where waste is separated.

Most sewage in England and Wales is treated, so it becomes sewage sludge. Dumping sewage sludge at sea was banned in 1998, so now it's either sent to farms (who use it as a land fertiliser), sent to be incinerated or sent to landfill.

Many areas of the country now have less than 10 years of landfill space left and the rubbish we produce is rising at least 3% a year nationally.

If every home recycled 50% of its rubbish the UK's annual CO_2 emissions would fall by six million tonnes.

RECYCLING – THE KNOW-HOW

Sorting out materials from our household waste ready for recycling helps in many ways because the less rubbish we send to landfill sites or to incinerators, the more valuable materials and energy we save. Unlike burying and burning rubbish, recycling allows materials to be used again and made into other useful products. For example, office furniture can be made from recycled plastic, sleeping bags are produced using shredded telephone directories and that fleece you're wearing could well be made from a batch of recycled plastic drinks bottles!

Most people think recycling is the hardest part of any lifestyle change, but it's actually simple – you just need to open your eyes and roll your sleeves up a bit.

KNOW YOUR RUBBISH TERMS

Recyclable – many products made from glass, metal, paper or plastic can be recycled and you can take them to your local authority's recycling centres – but make sure you put the right materials in the right bins (e.g. no green bottles in the brown bottle bank). Many of the large supermarkets, or main shopping centres provide recycling banks so you could use those facilities when you're getting your weekly shopping.

Recycled content – some products are labelled as 'recycled' even when the amount of recycled content is quite low. And sometimes it's not clear if this claim applies to the product/content or the packaging. A non-misleading and clear claim will tell you what percentage of the product/content (and/or packaging) is recycled or recyclable.

Biodegradable – means that a product will break down naturally in soil or in water. But products/contents and packaging take many years to break down and release harmful substances or gases in the process. Again, check that the claim really explains what is meant by biodegradable as most everyday products biodegrade eventually – but if that means in 200 years time, then that's not really acceptable!

WHAT TO RECYCLE

PAPER AND CARDBOARD

Recycling paper reduces pressure on natural resources and uses 30–70% less energy than producing paper from virgin materials. There should be a paper bank recycling facility near you – but make sure you put the right type of paper in the right recycling bank. Some, for instance, don't want 'coated' papers that are used in magazines and on some types of packaging.

Recycled products can contain varying amounts of recycled materials. Recycled papers, for example, have different recycled contents and grades of waste paper used. A paper product isn't considered to be recycled unless it contains at least 51% post-consumer waste, so products with 100% post-consumer waste content are the most environmentally-friendly. Recycled paper can be used for all sorts including toilet roll, cat litter, kitchen towels, stationery and animal bedding.

METALS

This is simple – wash and recycle all your tins and cans. Don't forget your aerosol cans are recyclable, just like any other steel or aluminium container.

You never know where your can may end up. A company called RE produce all kinds of stuff from recycled metal including picture frames made from old oil drums and bowls made from recycled telephone wire and newspaper. (www.re-foundobjects.com)

PLASTICS

Visit www.recoup.org for more details on different types of plastic and where you can recycle them. They also give you details on how to buy recycled plastic products. We still have a long way to go with recycling plastic in the UK; most of it is exported to China, but more and more companies are realising that there is profit in plastic waste.

ELECTRONIC EQUIPMENT

Over 1.5 million computers are dumped in landfill every year in the UK. Soon the WEEE Directive will mean that no electronic equipment can be disposed of in landfill, which will reduce pollution and cut the volume of waste going to landfill. However, many local authorities provide a free service for households, so give them a call first to see what's on offer. Try Action Aid's National Recycling Scheme for printer cartridges and mobile phones. Some Oxfam shops also take back your old mobiles. (www.nru.org.uk)

If you've really finished with that computer, send it to be refurbished or resold. Many international charities or schools can benefit from your unwanted equipment. Try Computers for Charity to start with. (www.computersforcharity.org.uk)

If your TV's on the blink or your stereo's stopped working, think about repairing it before you buy a new one; it's often cheaper, and cuts down on waste too.

BATTERIES

Many older types of battery contain potentially harmful metals such as mercury and cadmium, and there are so many different sizes and types of batteries that sorting and recycling is difficult. Rechargeable nickel cadmium batteries, found in drills or mobile phones, can be recycled – visit www.rebat.com for more details. Other types of battery are more tricky, so try not to use them.

BIG STUFF

Some recycling needs a bit more effort, but you'll be making a huge difference in helping to reduce your waste. Recycling can go much further than the contents of your dustbin.

For example, if you're replacing your fridge, freezer, washing machine or any other appliance with a nice new energy-efficient one, ask your local council if they can collect your old one and recycle it.

When it comes to furnishings, if you're anything like me, then a bit of the old designer mojo never goes amiss. We're funky chicks at heart so we do enjoy checking out and lusting over nice bits of furniture. After all, we want our home to scream 'style darlings' rather than it being overtly eco-freako!

In the past, my husband Gil and I, have taken the furniture store drug and bought a chest of drawers or some groovy kitchenalia, but we've come to realise how important it is to try and buy second-hand, good-quality furniture. Over the years we've managed to retrieve and revitalise a beautiful five-door carved wardrobe, three chests of drawers, and a massive table and cabinet made from driftwood. Other bits and bobs around the house include another Victorian pine table, a couple of armchairs donated by a friend and a futon dragged from a skip by our mate Nigel!

If you really don't want a bit of furniture any more and it's still in good nick, you should see if you can get it collected. You can also donate furniture in good condition to a charity or your local sports club for use in the clubhouse – some may collect it for free. Also, the Furniture Re-use Network (FRN) co-ordinates over 300 projects in the UK. (www.frn.org.uk)

If you've got loads of stuff to clear out, then why not take it to a car-boot sale and make yourself a few quid in the bargain? There are also loads of local charities who can find a good home for all kinds of clothes and household items that you may no longer need or want.

GARDEN AND KITCHEN WASTE

Instead of binning it, try composting your garden and veggie peelings. It's an easy process and not only will your dustbin be less smelly, but you'll be improving your local environment and even saving money in the bargain. If you don't have a garden to compost in, some local authorities have set up community composting sites. Visit the Community Composting Network (www.othas.org.uk/ccn) for more information.

GLASS

Glass is one of the best materials for recycling, as it can be recycled again and again, saving energy and raw materials. Before recycling your glass, make sure you wash the bottles and jars, and remove any tops or plastic attachments. Do try and put your glass in the right coloured banks – any contamination will lower the value of the recycled glass.

Why not check out Milagros and their fascinating range of recycled glassware for inspiration. (www.milagros.co.uk)

HOW TO RECYCLE

BANK SCHEMES

Most local authorities now provide local recycling collection points for glass, newspapers and magazines. Some also provide banks for plastics, textiles, books or aluminium foil. Visit www.recycle-more.co.uk to find out what banks are in your area.

KERBSIDE COLLECTIONS

In addition, many local authorities operate door-to-door (or 'kerbside') collection schemes for some items such as tin cans. Promotion schemes for domestic composting are also growing in popularity. Contact your local authority for further information.

CIVIC AMENITY SITES

Civic Amenity Sites were first provided as a separate facility from landfill sites, to allow the public free disposal of bulky domestic waste. These sites are now operated on behalf of county or unitary district waste disposal authorities, and are often called Household Waste Recycling Centres. For details of your nearest Civic Amenity Site and details of all the recycling services in your area contact your local council's recycling officer. For all household waste and recycling enquiries call the Waste Watch Wasteline (Mon–Fri, 10 a.m.–5 p.m.) on 0870 243 0136 or visit their website www.wastewatch.org.uk for more information.

- Waste Connect is a handy database which lists every recycling facility in the country. Simply type in your postcode to find what's available in your neck of the woods. (www.wastepoint.co.uk)

- Waste Watch provides information about what and where you can recycle. It also produces useful fact sheets on materials such as plastics and batteries that can be tricky to find recycling facilities for. (www.wastewatch.org.uk)

- Recycle More shows you how to increase recycling rates in your home and provides more information on recycling labels and what they mean. (www.recycle-more.co.uk)

- The website www.wasteonline.org.uk also gives you loads more information on recycling in general.

LETTERBOX DETOX

Just think. Up to 20% of what comes through your letterbox is never opened and up to 60% is never read. If we can stop it coming through our letterboxes in the first place then we can avoid all the waste and pollution it takes to produce and distribute it – and then to dispose of it.

The Mailing Preference Service (www.mpsonline.org.uk) can remove your name from all direct-mailing lists, but they can't remove you from mailing lists of companies you have actually bought things from or charities you have donated to. To stop receiving this mail you will have to contact the relevant companies directly.

WHERE TO BUY RECYCLED PRODUCTS

- Waste Watch's UK recycled products guide tells you where you can buy recycled goods of all kinds. (www.recycledproducts.org.uk or Tel: 0870 243 0136)

- Natural Collection Catalogue has a wide range of environmentally friendly and recycled products. Friends of the Earth receive a donation from sales. (www.naturalcollection.com or Tel: 0870 331 3333)

- Green Choices provides a mass of information for those wishing to make green choices including details of many recycled products for the home. (www.greenchoices.org)

GET PREPARED WITH SOME NEW TOYS

Crush-a-Can

Show your support for the environment by using your feet. You can buy a dual Crush-a-Can from Lakeland (www.lakelandlimited.com) that easily flattens bulky aluminium and steel cans safely and ready for recycling. The Crush-a-Can takes cans up to 475 ml and costs around £12.95.

A 'Rubbish' Invention

Made of strong, durable, laminated board, these eye-catching boxes are both smart enough to be on show in your home and easily carried to the recycling banks. As they have a wipe-clean finish, you can keep up the good work by using them over and over again! Each holds up to 16 bottles, 40 newspapers or 40 tins/cans. A pack of three – one of each design – costs approximately £1.95 from Lakeland.

WHAT DO ALL THE LOGOS MEAN?

It's not always easy to find out if one product really is more 'green' than another. A lot of products carry environmental claims and labels. Whilst some of these give useful information on what the product is made from, or the best way to use it or dispose of it, others are vague so don't have any real value to us as purchasers. The following should help you to decide what's 'puff' and what's 'buff'.

MOBIUS

The Mobius loop is probably the symbol we all think of when anyone mentions recycling. Essentially, it shows that something is capable of being recycled which doesn't necessarily mean that the object itself has been made from recycled material, so it can be misleading if more information isn't provided on the product label.

In this example, the use of the Mobius loop is more meaningful, because it shows that a particular product or object contains a percentage of recycled material. However, the use of this symbol is voluntary.

THE EU ENERGY LABEL

Manufacturers and retailers must, by law, tell you about the energy efficiency of a range of electrical appliances such as washing machines, dishwashers and fridges. Products are rated from 'A' to 'G' with 'A' being the most efficient and 'G' being the least efficient.

ENERGY EFFICIENCY LOGO

Energy Efficiency is a government-backed initiative run by the Energy Saving Trust and is aimed at raising awareness of the benefits of energy-efficient products that save you money and energy, and help to protect the environment. The Energy Saving Trust's blue and orange Energy Efficiency logo appears on a range of products – from light bulbs to laundry appliances – and indicates the most energy-efficient appliances on the market. (www.saveenergy.co.uk)

THE EUROPEAN UNION ECOLABEL

This label can be found on a wide range of household goods, including kitchen towels, toilet rolls, washing powder and paint. It shows that a product has met high environmental standards as the manufacturer will have had to apply to use the symbol on their products.

GLASS

This symbol shows that glass can be recycled – but make sure that when you take your bottles to the bottle-banks you put the right coloured glass in the right banks. It's too easy to mix up green and brown bottles, but the cost of sorting out these mistakes can be huge.

METALS

Recyclable aluminium

recyclable steel **Recyclable steel**

PAPER

National Association of Paper Merchants

To be given this mark, paper or board must be made from a minimum of 75% genuine waste paper or board fibre, no part of which should contain mill-produced waste fibre.

WOOD

The Forestry Stewardship Council (FSC) logo identifies products which contain wood from well-managed forests which are independently certified in accordance with the rules of the FSC. FSC Trademark 1996 Forest Stewardship Council A.C.

LOW VOC CONTENT

VOCs are volatile organic compounds and are found in many paints and household cleaning agents. Look for labels that show a minimum or low VOC content. Use water-based paints where possible. Contact the Association for Conscious Building on 01559 370908 for information on non-toxic paints.

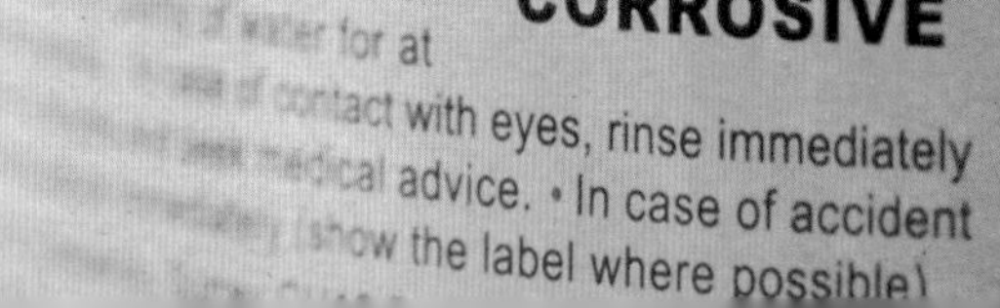
BLOCKER
products,
may be
children.
place and
protection.
After
water for at
CORROSIVE
contact with eyes, rinse immediately
medical advice. • In case of accident
(show the label where possible)

PLASTICS

There are various types of plastic which are identified as follows:

- **Polyethylene Terephthalate (PET or PETP)**
  Found on the bottom of many plastic bottles, containers for food products and carbonated soft drinks.

- **High Density Polyethylene (HDPE)**
  Found in containers for milk, household cleaning products and containers for cosmetics, shampoos and deodorants.

- **Poly Vinyl Chloride (PVC)**
  Found in a wide range of consumer products such as cling film, bottles, credit cards, audio records and immitation leather, as well as materials such as window frames, cables, pipes and flooring. It is also used in car interiors and by hospitals for medical disposables. The production of PVC creates and releases one of the most toxic chemicals, dioxin.

- **Polypropylene**
  Found in clothes, bottles, containers, caps and moisture-proof wrapping.

- **Polycarbonate**
  Found in plastic baby bottles, beakers and tableware.

- **Chlorofluorocarbons (CFCs)**
  Used to produce expanded polystyrene, but alternatives are being sought.

- **Polystyrene**
  Found in kitchen utensils or, in an expanded form, in insulation and ceiling tiles. It is produced in three main forms:

  Crystal polystyrene (GPPS)
  Clear and brittle material used to make CD jewel cases, bottles for pills, tablets, capsules and thin-walled cups.

  High-impact polystyrene (PS-HI)
  Used in thermo-formed containers for dairy products.

  Foamed polystyrene (PS-E)
  Used in clamshell containers (e.g. for fast food), meat trays and egg cartons. It is also used in loose-fill and moulded protective packaging for shipping electronic goods and other fragile items.

If you're still not sure what to look for you can find more information on the Department for Environment, Food and Rural Affairs (Defra) website, www.defra.gov.uk, including a copy of the Green Claims Code and more information on the European Union ecolabel and EU energy label.

What you have saved . . .

WHAT YOU HAVE SAVED . . .		SO TREAT YOURSELF . . .
Space in your kitchen. All those used bottles and jars can take up masses of floor and counter space.	→	Put fresh flowers, fruit and vegetables in the space instead.
Cost of antique Victorian wardrobe = £1,500; restoration cost = £500 for a second-hand wardrobe plus £50 for varnishes and other materials. Saving = £950.	→	Find and restore that four-poster bed you've always dreamed of.
Faith in your local council. It doesn't just have to be you doing all the hard work; find out what your local council and other organisations can do for you and take advantage!	→	Feel encouraged to be more involved. Join a local community project and give something back.

If you just do one thing . . .

Open your eyes – is the world the same place it was ten years ago? Reduce the damage you cause; reuse the things you throw away that others would kill for; and recycle – it's easy when you know what to look for.

CHAPTER THREE

Home truths

The British public wastes around £5 billion worth of energy a year.

During the golden years of gas and oil exploration in the UK we were self-sufficient in gas and even exported it. It is sobering to think that within the next 20 years we could be importing up to 80% of our gas from places as far flung as Malaysia.

If we think gas is expensive now, what will it cost then?

The key to reducing energy waste in the home is to ensure that we make the most efficient use of the energy we consume so that we cut our monthly fuel bills but still retain the same level of comfort that we're used to now.

Every year, just over a quarter of the UK's carbon dioxide emissions (which are known to cause climate change) originate from the energy we use to heat and power our homes.

Almost all our energy comes from fossil fuels (oil, gas, coal) and burning fossil fuels releases harmful gases like carbon dioxide into the atmosphere. This, in turn, causes changes in our climate.

By being energy efficient in terms of the way we power, heat and light our homes we can reduce our energy needs and reduce our reliance on power stations. That has the knock-on effect of cutting down our contribution to climate change. There are lots of things you can do to save energy in the home and many of them will mean huge savings on your fuel bills too. But, first you need to start to recognise your own energy consumption and work out what you could do to be less wasteful – the benefits are that you'll be saving money on your fuel bills and doing your bit for the environment too.

Like any lifestyle overhaul, some of these actions require a bit of thought. However, once you've made an effort for a few weeks you'll find that it becomes second nature and you don't even have to think about it. Being energy efficient can be one of the biggest money-savers in the home and, if you're not convinced after following the advice in this book, just you wait until you get your next monthly fuel bill!

Remember, nothing in this book is about deprivation. It's simply about making sensible and more responsible choices, rather than going without everything.

HOME ENERGY AUDIT

To work out how you use energy in your home and how you might be wasting money, it helps if you do a home energy audit. That way you can get an overall picture of where the biggest areas of waste are and what you can do to remedy things. Answer the following questions to get yourself thinking. You should try to keep a permanent log of all your home and its current energy-saving methods.

What type of home is it? (detached, flat, bungalow, etc.)	
How many floors does it have? (excluding loft and cellar)	
Do you have a loft and is it insulated? If 'yes', how thick is the insulation?	
Do you have cavity-wall insulation?	
What sort of windows do you have?	
Do you have any draught-proofing?	
What sort of heating system does your home have?	
What sort of heating controls do you have?	
How old is your boiler?	
Do you have any hot-water controls?	
Is the hot-water tank insulated?	
Are the hot-water taps lagged?	
Do you have any low-energy light bulbs in your home? If 'yes', how many?	
In total, how many of the following appliances do you have? fridge/freezer / fridge / freezer / washing machine / dishwasher / tumble dryer	

(Based on the Energy Saving Trust's online Home Energy Audit)

Now you have this information, you can begin to ask yourself where you can make the biggest savings. The information that follows in this chapter will help you to make positive changes and help you see why having a boiler over 15 years old, for example, is costing you money. Every inch of your home has money-saving potential – you just need to know what is being wasted where and how to fix it.

PAYBACK TIME

Some methods of energy saving are fairly cheap and you can do them yourself – for example, draught-proofing windows and doors or fixing a jacket to the hot-water tank. Others, such as double glazing and cavity-wall insulation, cost a great deal more and you will need professional installers.

When you're thinking about being more energy efficient in your home you need to consider the payback period – this is the amount of time the initial outlay will cost you to recoup. Although double glazing and cavity-wall insulation will significantly cut heat loss, it could still take a lot of time for the costs of what you've paid out to be regained through cheaper fuel bills.

Payback can be measured like this . . .

$$\frac{\text{The cost of work (in pounds)}}{\text{Fuel savings per year (in pounds)}} = \text{Payback period (in years)}$$

ENERGY ADVICE CENTRES

- Energy Advice Centres provide free, expert and impartial advice on how to save energy and money.
- They can provide you with a free DIY Home Energy Check, which allows them to assess what the most cost-effective ways would be for you to save energy in your home.
- They'll also tell you what grants are available in your area and provide details of your nearest approved professional installers, if you need one. In addition they can provide general information and tips on energy efficiency over the telephone, face-to-face or through their leaflets and information packs.
- Call the EEAC freephone hotline – 0800 512012 – during office hours. For full list of UK EEACs refer to Energy Saving Trust website. (www.est.org.uk)

DON'T BE DAFT – STOP THAT DRAUGHT!

Draughts are a clear sign that your home's losing precious heat – which means that the money you're spending on heating your home is wasted money. Draught-proofing is one of the very first things you should do and will improve the comfort level of your home. It's a job that most of you could tackle yourselves. Draught-proofing strips are relatively cheap to buy and can be picked up from any good DIY store. However, make sure that you leave adequate ventilation to prevent condensation and mildew.

It's also vital that gas appliances get a good supply of fresh air. They need this to burn efficiently and to take away harmful fumes. Whatever you do don't go blocking up air vents or grills in the walls. There must be adequate ventilation in every room where there are any heating appliances such as boilers or open-flue gas water heaters.

In 'wet' areas such as kitchens and bathrooms, think about installing heat exhange units with an in-built humidistat. These cut heat costs by reducing moisture and extracting warm air and returning it to the room. Contact Ventaxia. (www.vent-axia.com)

The percentage levels below are dependent on several factors such as when your home was built. It's unlikely that your home will leak the maximum percentage on each area of loss, but there will be weak areas that you could address.

Total heat loss in a typical house

Windows – up to 10%
Walls – up to 35%
Floor – up to 15%
Roof – up to 25%
Draughts – up to 15%

WINDOWS AND DOORS

Heat lost from windows and doors can be tackled in two ways: by making sure that the doors and windows are not left open needlessly, and by fitting appropriate draught-proofing.

Fitting draught brushes to the bottom of doors and adding seals to the door frames will help to reduce draughts. The letterbox can also be fitted with a cover – usually two nylon brushes – and draughty keyholes can be fitted with a cover.

The amount of heat lost through windows depends on what type of frames you have and whether or not your windows have any kind of secondary glazing. Metal window frames, for instance, are really good at conducting heat, which means that a lot of heat is lost through them. Wooden frames, on the other hand, are poor heat conductors, so heat tends to stays in the room longer. The older the property, the more likely you are to have windows that don't fit properly and cold air can blow in through the gaps.

The cheapest and simplest way to 'double glaze' a window is to attach some form of plastic sheeting over it. Most DIY shops sell kits for a reasonable price and range from cling-film-type materials to more long-lasting plastic sheeting.

The cling-film stuff comes in large widths that can be cut to size depending on the shape and size of your windows. It's a bit fiddly, but once you've done one you'll soon get the hang of it. Basically, you attach double-sided tape to the inner face of the window frame and then you use a hairdryer to stretch the cling film taut. It's very inexpensive, but it will only last for one winter season and you won't be able to open the windows you've sealed.

One alternative is to buy flexible plastic sheeting, which is still fairly cheap, but can be taken down and reused each winter. Again, it's important to check that there is adequate ventilation left in the room so that you can still open at least one window.

Finally, windows nearly always need some sort of covering – curtains being the most obvious choice for most. It's worthwhile investing in a pair of well-fitted (and preferably lined) curtains. Thermal lining can be bought from most good fabric stores and only costs a few pounds a metre. Try to choose thick fabric that will keep the heat in and make sure you go for fullness, so when you close your curtains there are lots of folds in them – the gaps help to trap warm air. Draw your curtains at dusk and open them first thing in the morning to make the most of 'passive solar gain' which means using the sun to heat your room, without costing a penny.

POTENTIAL SAVING

Draught-proofing windows and doors
Cost = £45–50
Saving = £10–20 (per annum)

Secondary glazing
Cost = £40–600
Saving = £15–25 (per annum)

Double glazing
Cost = £2,000+
Saving = £20–35 (per annum)

WALLS

Most heat is lost through the walls of a house, but this can be remedied in a number of ways and can reduce heat loss by two-thirds. If you have cavity walls, they can usually be filled with a special insulating material, but this needs to be done by professionals and the costs are quite high – although some grants may be available.

If you have solid walls, then you could go for some type of internal insulation – for example, some kind of insulated board fixed to the wall. The drawback is that any kind of wall insulation costs money and for internal walls it means that the rooms will need to be redecorated too. At my home I have installed external insulation on the side and rear walls. It cost around £15 for for 200 square metres but it means I only have to use a fifth of the heat that I used before.

POTENTIAL SAVING

Cavity wall insulation
Cost = £300–400
Saving = £60–70 (per annum)
Internal wall insulation
Cost = £700–2,000
Saving = £40–60 (per annum)

FLOORS

You can lose as much as 10% of heat through un-insulated floors. Heat escapes as a result of gaps in floorboards, or near outside walls where heat escapes more quickly because the floor is in contact with colder earth at the sides of the house.

All those trendy polished floorboards we favour so much today can account for 15% of heat loss. The easiest way to reduce this is to cover the floor with underlay, carpet, mats, woodblocks or cork tiles.

ROOFS

As much as 25% of your heating bill can be saved by making sure that you've got good loft insulation. Lofts need to be insulated to a depth of 250 mm – a high proportion of the heat that escapes from our homes goes through our roofs.

Two materials can be used to insulate loft spaces – the most common is a roll of insulating material, which you can lay yourself. Alternatively, loose filled materials can be poured between the joists to an even depth.

Although loft insulation is generally seen to be one of the easiest DIY jobs, it's important to ensure you install it correctly to achieve the full potential. So, if you're going to tackle this yourself make sure you get the right tools for the job. You must wear protective clothing (goggles, a face mask and gloves), as insulation materials can be very irritating to the eyes, skin and throat. Any DIY store can provide you with this equipment as well as leaflets that explain how to do the job.

POTENTIAL SAVING

Loft insulation
Cost = £150–300
Savings = £60–80 (per annum)

POTENTIAL SAVING

Floors
Savings = Depending on the materials, you could save £200 (per annum)

BRIGHT IDEAS

About three-quarters of all domestic energy wastage could be eliminated simply by installing long-lasting, highly efficient light bulbs. If every UK household installed three energy-saving light bulbs we would save enough energy each year to run all of the country's street lighting. They may cost a bit more than standard bulbs (usually around a fiver) but they last twelve times longer and for each one you fit, you can save up to £7 on your annual electricity bill.

We're a nation obsessed with interior design, and the average household lighting bill has rocketed. Long gone are the days when our living room was lit with a single bulb dangling from the centre of the ceiling. These days our living rooms are full of up-lighters, spotlights and halogens and wall lights. Overall Britain's domestic lighting bill is now about £1.2 billion a year.

Energy-saving bulbs are particular good in areas where you have a light on for a long time. For example, if children won't go to sleep unless there's a light left on in the hall use an energy-saving one. Energy-saving lighting comes in all shapes and sizes now, and are much more effective than the original energy-saving bulbs, so you can be design conscious and less wasteful at the same time.

ROOM WITH A PHEW!

We need to make sure that we're using heating as efficiently as we possibly can. Some heating systems are more controllable than others, but the aim is the same: heat the rooms you want to, at the most suitable temperature, at the times of day you need to – and as cheaply as possible.

It's generally recommended that room temperatures should be somewhere between 18–21°C. Keeping the temperature at a reasonable level means that heat is lost more slowly to the outside air. It's even better if you're able to adjust your heating appliances to give you what you need, when you need it.

Room thermostats are usually placed on the living room wall or in the hallway and they respond to the temperature in that room. When the room's warm enough the thermostat sends a signal to the central-heating pump to stop. That means water is no longer circulated around the radiators.

Some thermostats can be fitted to gas fires or electric heaters, or you can install thermostatic radiator values, which allow you to control the temperature in rooms other than the one in which the main thermostat is placed.

As a general rule, only occupied rooms need to be kept warm. Although some heat is needed in other rooms to prevent condensation and damp building up, this heat can be kept fairly low.

If you want your heating to come on when you first get up to make your morning cuppa and then again before you get in from work, the best way to do it is by having an automatically controlled timer system installed. This controls the times at which the heating and hot-water systems come on and off. The simplest models have a time switch, whereas the most sophisticated allow different programming of heat and hot water for each day of the week.

Heat the rooms you want to, at the most suitable temperature, at the times of day you need to as cheaply as possible.

OLD BOILERS

- **Replace boilers 15 years old or older** because it could save you over 40% on your fuel bills.
- **Condensing boilers** are the most energy-efficient type of boiler on the market and could save you around a third on your bills straight away. And if you plump for a modern heating-control system at the same time, then you could save even more. Visit www.boilers.org.uk to compare old and new boilers.
- **Insulating a hot-water tank** or cylinder should be a top priority as un-insulated tanks lose a lot of heat. An insulating jacket only costs a few quid from a DIY store and pays for itself within a couple of months. If you go for one that's 75 mm thick you could save up to £15 a year. Jackets come in a range of standard sizes and are fitted with belts that fasten snugly around the tank, so it's something you can do yourselves. The effect of insulation will be to reduce the rate at which heat flows out of the tank, so less heat will be needed to keep the water at a reasonable temperature.
- **Hot-water pipes** should also be insulated to reduce heat loss. You can use lengths of rigid foam tubing, which you cut to size and then simply clip into place around the pipes. Again, this costs very little and can be done quite easily.

RADICAL RADIATORS

- **Foil behind your radiators** If you put foil behind your radiators that are against outside walls this will help to reflect heat back into your rooms. You may even be able to turn down the thermostat a touch and make even more savings on your bills. Normal kitchen foil pasted to a board will do, or you can buy specially designed foil from a DIY store for about a tenner.
- **Putting shelves up above your radiators** may sound a bit barmy, but it means that you're saving as much heat as possible because it's deflected back into the room. That can save you around five to ten pounds a year. The shelf should be fitted a couple of inches above the radiator for it to have maximum effect. It also helps if you move furniture away from radiators too, as furniture absorbs heat and stops it warming up the room.

ELECTRIC SHOCK TREATMENT

We spend £700 million every year on electricity to power a whole host of electric household gadgets – which means that we're demanding more and more energy from power stations. One of the best things we can do to offset the damage power stations cause to the environment is to switch to a 'green' energy supplier.

Green energy is energy generated by natural resources such as the sun, wind and water. Nearly all the major utility companies provide a green tariff of some kind, so you may not even need to switch your current supplier.

You don't need any new cables or meters, as green electricity is delivered like normal electricity through the national grid. The difference is that your supplier is buying a percentage of electricity from renewable sources. This electricity is then fed back into the national grid, replacing traditional 'brown' electricity.

Look into the costs to see if it can work for you. There are two types of green tariff. With a renewable tariff, every unit of electricity bought by you is generated from renewable energy sources. Or, if you opt for an eco-funds tariff, which is an additional premium, this is invested in a fund and used in new renewable projects.

The more green electricity we purchase, the cleaner the supply coming through the national grid will be.

Check green prices and what's available on www.greenpower.co.uk or the Green Electricity Marketplace provides information on all the green tariffs available. (www.greenelectricity.org)

CRAZY COOKERS

When cooking, make sure you choose the right pan size for the food you're cooking. If you cut food into smaller chunks and put lids on pans it will cook much quicker too, saving more energy. If you don't know how to cook – why not learn? Make your own ready-meals by doubling up on quantities and freezing one half for a later date.

MICROWAVE MADNESS

I'm personally not a fan of what I call 'ping cuisine'. Still, microwaves are a fairly energy-efficient method of cooking. They do, however, destroy much of the nutritional value of our food – so it's effectively 'dead' food we're shoving into our bodies.

Most of the stuff we tend to bung in our microwaves are pre-packaged ready-meals. A wasteful 13–16% of every pound spent on packaged food goes on the packaging alone. On a £10 shopping bill, that could be as much as £1.60 spent on stuff you can't eat – you are effectively paying for disposal of the packaging.

FRIDGES AND FREEZERS

- Fridges and freezers are the real workhorses in our homes and, unlike other appliances, you can't switch them off every day. In Britain, we use around £1.2 billion-worth of electricity just running our fridges and freezers each day.

- A high proportion of that electricity is wasted, because many of us are using old fridges and freezers that use more power than is necessary. In fact, they may be costing us more in electricity bills than the price of a new one. An energy-efficient fridge-freezer uses nearly a third of the energy to do the same job as a ten-year old appliance. That could amount to a saving of £45 a year on your fuel bill.

- You can also cut costs by not leaving the fridge or freezer door open longer than you need to – it takes three to four minutes for it to regain its temperature, which means you're wasting energy unnecessarily. And the same goes for putting hot food into the fridge – you're just making it work harder than it needs to.

- Try to defrost your freezer regularly and check that the door seals are still working properly. Regularly defrosting your freezer and keeping it packed full to avoid wasting energy will save you money. You should also check the seals on your fridge and freezer to see if any warm air is getting in. The seals should be tight enough to hold a piece of paper securely when the doors are closed.

WASHING MACHINES AND DISHWASHERS

- Washing machines and dishwashers use loads of water, so try to put them on only when they're full. That way you make the most of the water and electricity they use. If that's not possible then most machines have half-load or 'eco-wash' buttons, so you should try to use those wherever possible.

- It's also worth looking out for the energy label when buying a new washing machine or dishwasher, as they'll show you whether the appliance is water and energy efficient – the water consumption should be less than 55 litres per cycle. A washing machine uses on average 80 litres of water, which is five times more than a large sink.

- Only wash clothes when they're dirty. If you've only worn an item once and it's still spotless and doesn't smell, hang it outside or in the bathroom when you shower to freshen it up. This saves time, energy, water, and prolongs the life of the garment too.

- If you go for a 40°C wash cycle rather than 60°C you'll be using a third less electricity. Reduce the wash to a 30°C cycle and the amount of electricity and money saved will be even higher. Unless you have very dirty and heavily-soiled clothes you really don't need a high-temperature wash.

- An energy-efficient dishwasher will cut energy wastage by half. When using your dishwasher, like your washing machine, it's best to wait for a full load or use the economy cycle if your model has one.

- If you really have to buy a tumble dryer, make sure you buy the most energy-efficient model you can as it could cut energy wastage by almost a third. Avoid trying to dry really wet clothes – wring them out by hand or spin them dry first. Not only will the clothes dry faster, but you'll save money.

POTENTIAL SAVING

By simply changing the way you use electronic products you could cut your energy bills by a quarter!

LOGOS A GO-GO

Look out for the Energy Label on fridges, freezers, washing machines, tumble dryers and dishwashers. They will explain what you're getting for your money in terms of how much the appliance costs to run, how well it performs, how much water it uses and even how noisy it is.

The more efficient the appliance, the less energy it needs to do the same job, so the more you get for your money. A-rated appliances are the most energy efficient with G-rated being the least efficient. Use the main A–G scale to find the best buy.

The label also tells you how much electricity the appliance used in standard tests. You can use this information to work out how much you might save if you choose different types of models. Actual savings will, of course, depend on how you use the appliance and how much you pay for your electricity.

WISHY-WASHY WASTERS

On average, a person in a developed country such as the UK uses about 150–160 litres of water a day and we tend to use it in the home in the following ways:

ACTIVITY	AVERAGE AMOUNT OF WATER USED IN LITRES
Taking a bath	120 litres
Running taps	1 litre
Washing	8 litres
Brushing our teeth	2 litres
Taking a shower	30 litres
Flushing the toilet	40 litres
Washing clothes in a washing machine	120 litres
Using a dishwasher	60 litres
Watering the garden using a sprinkler	60 litres
Washing the car with a hose pipe	90 litres

We really do take water for granted. After all, we simply turn on a tap and out it comes.

The Thames Water region uses 55% of available rainfall. Another 800,000 people will be living in London by 2016 – the equivalent of the population of Leeds. We will need to find an extra 128 million litres of water per day to supply them.

We really do take water for granted. After all, we simply turn on a tap and out it comes. However, only 3% of the water on our planet is fresh water – the other 97% is undrinkable sea water. What's more 80% of that 3% of fresh water is locked up in ice caps lying deep underground. This means that approximately only 1% of the planet's fresh water is drinkable.

Every time we use water it has to be cleaned. Most of the water we use ends up down a sewer and is then transferred to a sewage treatment works. With increased flooding, sewers can overflow and this raw sewage can find its way back into rivers and other water supplies.

Most waste will break down naturally, but sometimes when there's just too much for the sewers to cope with, pollution occurs. When sewage gets into the water supplies, bacteria naturally begins to break it down. However, this process requires oxygen and this means that oxygen is rapidly removed from the water supply.

WHAT YOU CAN DO TO SAVE WATER

DO

- Take a shower instead of a bath – a five-minute shower uses up about a third as much water as an average bath.
- Make sure taps are turned off fully. A dripping tap can waste as much as ten litres a day.
- Use environmentally friendly household cleaning products where you can. Or alternatively try using borax or vinegar to clean the bathroom.

DON'TS

- Avoid using more water than you need. When you're cleaning your teeth, don't leave the taps running.
- When you have a bath only run as much water as you need – if you fill a bath right up it can hold more than 120 litres of water.
- Don't use too much detergent when washing up and hold back on pouring in the powder to your washing machine. Even environmentally friendly detergents need to be treated to some extent.
- Never pour oil, paint or any other chemical liquid down the drain or toilet.
- Try not to use bleach or other disinfectant unless it's really unavoidable.

DON'T BE A DRIP!

Try to make sure you always turn taps off completely and try not to leave them running when cleaning your teeth – it wastes loads of water. A dripping tap can also waste a bathful of water every week, so make sure you fix any dripping taps in the home.

If you think there's a leak in your property (maybe you've had an unexpectedly high water bill) then report it to your water company. It's your responsibility to fix any leaks, but keep any plumber's receipts, as you may be able to claim a refund from your water supplier. If you're not sure who your water supplier is, www.water.org.uk will tell you.

You can save water in the kitchen, too, by using a bowl of water to wash fruit and vegetables instead of running them under the tap. And if you generally run the tap to make sure water's really cold before you drink it, why not fill up a jug and keep it in the fridge instead?

FLUSHING MONEY DOWN THE PAN

Toilets are huge water guzzlers, accounting for 35% of domestic water use. Old toilets use 9.5 litres each flush, although those installed since 1993 use only 7.5 litres. We don't need to use so much water when flushing, so here are some ideas to cut down:

- **Fit a water 'hippo' bag**, full plastic bottle or anything that will take up space in the cistern. 'Hippos' will save up to three litres of water with every flush. If one toilet is flushed ten times a day, this equates to a water saving of 25 litres per day – that's enough for a five-minute shower.

- **The eco-flush** is another device and is a surprisingly simple invention that offers the user a choice of flush with high and low settings. This could reduce water used to flush the toilet by a third. It takes a few days to get into the habit of selecting the right setting, but it's well worth it for both the financial savings (if your water is metered) and the environment. Installation is fairly simple too.

- **Don't use the toilet as a dustbin.** When sanitary products and condoms reach the sewage treatment plant, any blockages that occur mean that the sewage will eventually end up in rivers and the sea. Make sure that items like tampons, sanitary towels, disposable nappies, cotton buds, condoms and cigarettes are put in the bin instead of down the toilet.

SEXY SUDS

For those who want something a little more adventurous and amorous, how about sharing a bath to save water? Put on your favourite Barry White album – OK, that may not be to everyone's liking – light some scented candles, pour a glass of wine and sit back and relax together. It's a great way to wind down, spend some intimate time with your partner and de-stress after a hard day at work.

FIVE-MINUTE SHOWER

This is so simple. Switch it on; get wet; switch off; soap up and rinse. A power shower running for 20 mins uses more water than having a bath. Make sure you're being efficient in the way you shower and save money from going down the drain.

If we all switched off our TVs at the socket,
we could save enough energy
to power a small town.

TOP TIPS FOR SAVING ENERGY AND MONEY

Most of us are new to thinking green so if all this information seems too much to digest, pick one tip from each of the following lists that is easiest for you and start doing it today!

No cost

- Standby buttons – that evil little red eye hypnotises us, and for the sake of convenience we tend to leave our TVs, videos, PCs and printers 'on'. If we all switched off our TVs at the socket, we could save enough energy to power a small town. And why not switch off your telly and talk or play a game instead? Have a romantic night in and use your energy in a more interesting way!

- Turn your heating thermostat down by 1°C to a minimum of 20°C – you will hardly notice the difference and it will cut your bills by 10% a year. Wear jumpers in the winter and don't shock your body by going from summer temperatures in your house to cold air outside – it confuses your metabolism and lowers your resistance to bugs.

- Move furniture away from radiators – the foam insulation tends to trap the heat.

- Choose the right sized pan for the food and cooker, cut food into smaller pieces and put lids on pans, as the food will then cook a lot quicker. If you are defrosting food, or just warming things up, microwave ovens are ideal, as they use much less electricity than conventional ovens.

- Regularly defrost your freezer and keep it packed full to avoid wasting energy. Check the seals are still intact to ensure no warm air is getting in. They should be tight enough to hold a piece of paper in place without it falling out.

- Try to use full loads when using your washing machine and use the low temperature wash of 40°C. On some washing machines, only cold water is used to fill the machine when it is set to a low temperature, which means that there is no need to heat up the central-heating boiler.

- The sun is the most readily available source of heat there is – and the cheapest! So, make the most of it by opening internal doors of any rooms that get more sun than others and let the warm air travel though your home on a summer's day.

- Avoid using tumble dryers and radiators to dry your clothes – on a nice sunny day hang them outside instead and let them dry naturally.

- Fix any leaky taps, as even a small drip can use a lot of water over time.

- Wash your car the natural way – wait until it rains!

CASE STUDY

CASE STUDY – STUDENTS

Oliver-Kissick Jones – 21, Marianne Wilsen – 20, Rachel Cave – 21, Victoria Watts – 21, Adam Riymill – 23, Ollie Hickman – 21

If there's one group of people who should be even keener to save money on household bills, it's students. However, when I visited this student house in London I realised that it was full of young people who partied hard and thought little about the effects of their actions. They split the bills seven ways even though some of them were more wasteful than others and, because they all paid an equal share, they had fallen into bad habits.

Eco-warrior Ollie had the constant job of turning off the lights and trying to educate his housemates about how to use the thermostat. He was having to leave a constant barrage of reminder notes which created conflict within the house as he tried to enforce his Eco ways. When they were not studying (!) they were party fiends. With a set of decks powered by many, many extension plugs, they partied into the night using all the lights they could. They confessed that, with hangovers the next day, the last thing they would think of would be sorting through their rubbish – and that rubbish could be two full bin liners a day, 14 a week – nearly four times the national average.

With the heating left on for over 12 hours a day, the students enjoyed a boiling house without any consideration of their bills. Having only lived in the house for six months, they had a lot to learn about economising now – otherwise their loans and savings were going to get eaten up by bills. Unless they took on my lifestyle rules, they could become wasters for all of their adult lives.

Low cost

- Fit energy-saving light bulbs – they use about a quarter of the electricity that normal 60-watt bulbs use. That could save you around seven to ten quid a year – and you won't have to change the blighters for years either, as they last for ages. Just seven low-energy bulbs could literally last you a lifetime – a good-quality low-energy bulb can last up to 12 years. 12 x 7 = 84 years!

- Draught-proof doors and windows – gaps in door and window frames are responsible for a fifth of wasted heat.

- Insulate water tanks and pipes – fitting an insulating jacket to the hot-water tank will keep water hot and save up to £20 a year. Lagging pipes could save you an extra fiver.

- Insulate the loft – 250 mm of loft insulation will cut your heating bills.

- Buy energy-efficient appliances – a fridge is the biggest guzzler of power in the home. A new A-rated energy-efficient fridge will cost you less to run. Just think what you could spend the money on instead!

- Double-glaze those windows – it reduces noise and condensation and keeps the heat in, reducing your fuel bills.

- Install an energy-efficient boiler – a high-efficiency boiler could save up to 40% on your fuel bills. Power companies and local authorities often provide grants to help meet the extra cost of installation.

Turn your heating thermostat down.
18–21°C is comfortable
and if you get chilly, wear a jumper.

What you have saved . . .

WHAT YOU HAVE SAVED . . .		SO TREAT YOURSELF . . .
A possible saving of £500 per year by insulating your home properly.	→	Replace your boiler or invest in loft insulation, you'll save more in the long term.
630 litres of water a week by taking a shower instead of a bath.	→	Feel less guilty about buying the occasional bottle of organic wine.
Yet another argument . . . Hotter environments can make us more stressed and irritable.	→	Have a peaceful home life for once. Turn down your heating and cool that hothead down!

If you just do one thing . . .

Try the five-minute shower challenge!

Get wet, turn off, soap up and rinse. A 90 litre bath costs 25p whereas a five minute shower using 25 litres of water costs only 7p on a standard combi boiler.

CHAPTER FOUR

How you shop affects what you drop

A LOAD OF OLD RUBBISH

As we've seen already, getting rid of all our waste harms the environment and uses energy. Many landfill sites are nearly full to bursting and we're running out of suitable land for new sites. It's time to reflect on exactly what you buy and how you can use your purchasing power in a good way.

In the UK each person wastes an average of £430 worth of food every year and in 21 years' time, supermarkets will have provided us with enough plastic shopping bags to cover the whole of England. I just hate this waste on financial and moral grounds. With millions of people suffering from hunger, there has never been a more appropriate time to get conscious of the food we eat . . . and waste.

SHOP SMART

Bombarded by advertising, special offers, money off, super-saver deals and three-for-twos, we're slowly being seduced by the advertisers and marketeers. Shops and supermarkets are becoming increasingly more adept at making us buy things we don't actually need or use, which means that we're literally returning home with bags full of rubbish every time we go shopping.

Remember my Lifestyle Rules – reduce, reuse and recycle. When it comes to shopping I'd like you to start to think about 'precycling' too. All that means is stopping to think about what you buy because, by making sensible purchasing decisions, you can reduce your household waste. Just think, you could cut your garbage by as much as 20% simply by being more 'shop smart'.

PACK IT IN . . .

Take a good long look at what's in your shopping basket, as it's worth just checking the amount of packaging you've got. The cost of packaging is included in the price of each product, so essentially you're paying for junk. In fact, packaging can cost as much as 16p in every pound of your purchase. So, if something costs two quid, you'll actually be paying 32p for the privilege of bunging away the waste packaging that comes with it. So, always look for the product with the least packaging, and avoid multilayered products. Anything with loads of excessive packaging means that you're literally throwing money away!

Buy in bulk instead of individually packed items. For example, if you buy your fruit and veg from the supermarket then buy it loose rather than prepacked. You can also reduce your rubbish by bringing your own bags. Lots of supermarkets offer bag-recycling schemes now and some give you money back on your shopping for using your own bags. Or, if you really want to ratchet up your green credentials, you can buy a fair trade unbleached calico shopping bag made especially for the Green Shop. It measures 32 cm wide by 40 cm high and costs around £3.

THE SEVEN DEADLY SINS OF SHOPPING

Do you recognise your deadly sin from the list below? If so, follow my advice to become more shopping savvy.

1 Gluttony

You're the typical impulse buyer – 20 handbags in various colours and sizes, 18 pairs of identical trousers, 14 different types of shampoo and conditioner. And how do you manage to keep on spending so much? Easy, you're always in debt and everything's bunged straight on to a credit card. You need to learn to be more discriminating. Think about a capsule wardrobe – some stable basics that you can mix and match. You really don't need 20 white shirts – just keep the best ones that you simply can't bear to part with and make sure you wear them. If you don't you need items of clothing, bag them up and give them to a charity shop or to friends who will wear them.

2 Greed

You have an insatiable desire for designer labels and expensive jewellery and you simply refuse to go for the cheap option. You're the king and queen of bling! You're driven by fashion's latest fads and you simply have to have the latest model. So, you need to think about reducing, reusing and repairing stuff if you really can't face the thought of buying bargains. Think about what you love most – if it's designer labels then buy classics that won't date.

3 Lust

You're an uncontrollable pleasure seeker and are very high maintenance. If you're a girlie, then you'll have heaps of high heels, handbags and sexy undies. If you're a fella you'll forever be buying 'boy toys' – battery-operated beer openers, games consoles and electronically-powered toy cars. Girls and boys – you need to get a grip on your spending habits. All this stuff costs money and you're bound to be buying far too much of it. If you really can't curb your lust of the finer things in life, try to buy the best you can and stick to one 'luxury' item per year.

4 Wrath

You just hate sharing, don't you? You hoard loads and loads of stuff that you never use but woe betide anyone who pinches a squirt of your ten bottles of perfume. Your wardrobe is a real mix of styles and you're forever buying stuff that suits your current mood only to have a hissy fit and chuck it in the darkest recesses of your closet when you see that it makes your bum look big. Every time you buy something new, give something to charity that you no longer need or like. Your task is to clear the clutter before you even consider getting anything new.

5 Envy

If they've got one, you want one. And if you could only be three stones lighter, then you'd feel much more attractive. But, the truth is, it's hard to keep up with the Joneses and trying too hard makes you miserable. You need to stop being so hard on yourself. Don't wait until you've slimmed down to a size 10 – store all your 'thin' clothes in a bag, label them and set a date of six months. Promise yourself that if you're still not in them by then you'll let them go.

6 Sloth

Well, your wardrobe is stacked with stuff from the 1970s that you meant to get rid of, but never quite got around to. You need to sort out your wardrobe NOW! Did you know that 80% of people wear only 20% of the clothes they own? You need to bag them up and take it to a charity shop, or have a clothes-swap party with some friends.

7 Pride

You want to look fabulous all the time, so you're forever spending a fortune on expensive face creams, make-up and other beauty products to keep yourself in tip-top shape. If you're a bloke, then you're a bit of a Beckham – bathroom brimming with 'grooming' products. While it's good to take pride in your appearance, you can overdo the purchasing in the pursuit of perfection. Limit yourself to one or two essentials. Ladies, you really do not need 14 lip-glosses in various fruity flavours. If you buy make-up then ask an assistant at a beauty counter to help you choose what suits your face, age and skin. That way you won't be tempted to buy what you've seen in a magazine in the hope that it'll suit you. Make sure that the product contains natural and non-toxic ingredients, that it has not been tested on animals, that packaging is kept to an absolute minimum and that the company has an ethical policy towards its workers and operations.

Did you know that 80% of people wear only 20% of the clothes they own?

SELF-TEST CHECKLIST

The next time you come home from the shops, carry out this simple checklist. If you're bringing home more than four things in each category, you're bringing home way too much! Choose goods with the least amount of packaging – one layer is enough.

- PLASTIC SUPERMARKET CARRIER BAGS
- CARDBOARD PACKAGING
- PLASTIC WRAPPING MATERIALS
- PLASTIC BOTTLES
- POLYSTYRENE PACKAGING

In 21 years' time, supermarkets will have provided us with enough **plastic shopping bags** to cover the **whole of England.**

TOP TIPS FOR LOW-WASTE SUPERMARKET SHOPPING

Make a list and stick to it – only buy what you need and try to resist those special offers

Buy fruit and vegetables loose and not prepackaged – it's cheaper, healthier and cuts down on packaging that you have to cart home with you

Go for glass bottles and jars or tins – cartons and plastic can be difficult and expensive to recycle, so try to avoid buying them

Use reusable goods which last longer than single-use disposable items – razors and nappies, for example

Take your own shopping bag and avoid using the free plastic bags

Use supermarket recycling banks when you shop rather than making an extra trip – this saves on time and petrol

HAPPY SHOPPING

BUY PRODUCTS THAT LAST

Look for high-quality goods that are made to last. They may cost a bit more now, but they'll save you money and protect the environment in the long run. For instance, if you buy cloth nappies, towels and hankies they actually work out less expensive to use than disposable ones, as you don't have to constantly replace them or throw them away.

Buy non-perishable products such as household cleaners in bulk, as you'll only have one container to throw away – but alternatively make your own cleaning products using things in your kitchen cupboard.

Buy refills for the products you buy regularly like washing powders.

LEARN TO SHARE

Some items such as tools, party supplies, sports equipment, etc. could be shared with your family, friends or neighbours. If you have items you don't need, no longer use, or no longer like, why not donate them to a charity or give them to a friend. And don't forget to check out second-hand stores for furniture, household appliances and other items. You'll be surprised at what bargains you can dig up.

JARGON BUSTER

Post-consumer – made with recycled material that came from a community recycling programme. 100% post-consumer is the ultimate in recycling.

Pre-consumer – made with recycled material that came from leftovers in the manufacturing process.

REUSE IT, REPAIR IT OR RESTORE IT

Try to look after things and make them last. Instead of buying brand new stuff all the time, see if you can repair the old stuff that you really love and want to keep. Yellow Pages lists repair services for a whole host of household items. Think creatively and try to find new uses for items you might otherwise bin. Make it a personal goal to keep as much as possible out of the dustbin.

CLEAN UP YOUR ACT

Reduce your reliance on hazardous household products. Read the label and go for safer and more environmentally friendly alternatives. Or 'do as granny did' and use what's in the kitchen cupboard to clean your home.

BUY RECYCLED AND RECYCLABLE ITEMS

When you purchase products made from recycled materials, you close the recycling loop. Look for glass bottles or recycled paper products and containers and check out the symbols to see what you're getting. There are a whole host of household items that can be made from recycled materials. For example, glass jars made from recycled glass, clothing made from recycled plastic drinks bottles, and kitty-litter made from recycled newspaper. And if you buy items in recyclable containers – which means that you can use them over and over again – then you won't have to throw the containers away. That means you'll be able to refill them, which should help to save you money.

CASE STUDY

TIBBETT FAMILY

Marianne and John and their four daughters: Catherine 2, Genevieve 4, Francesca 6 and Annabel 8.

This family were chucking away about 20 bags of rubbish every week. We tipped 17 bags of their waste on their back garden and got the whole family to sort through it. We found they were throwing out huge amounts of food, in fact 20% of their rubbish was food waste. I explained to the Tibbetts that their local authority actually had one of the best recycling rates in the country (39%) and that they as a family were letting the side down. I also pointed out that they would soon have a doorstep recycling scheme put into place so they needed to learn how to separate their rubbish anyway. By the time we had completed the reduce, reuse and recycle lesson, we had got their landfill waste down to three bags and with practice and reduction of packaging, they will reduce it even further.

It was the amount of food they were wasting that I really wanted to get to grips with. Their food waste nearly filled a wheelbarrow and apart from the fact they were wasting a huge amount of money, they had been sending it to landfill which would result in methane being produced as it broke down.

We eat huge amounts of meat and therefore have millions of ruminants roaming the Earth eating grass, farting and producing damaging methane. Their poo also produces huge amounts of nitrogen that pollutes water sources. I did not suggest that the Tibbetts became vegetarians but suggested that they cut down on the number of meat-based meals – to save money and waste.

CHICKEN HEAVEN

The family have a good-sized garden and I came up with a solution that was perfect for them: chickens!

I contacted a company called Omlet who supply chickens complete with their own 'eglu' which has to be the most stylish and innovative chicken house ever. It is a great way to keep chickens as pets and because of its clever design, you can have fresh eggs whether you live in the country, city or suburbs. I loved it so much I am going to have one. Find out more at www.omlet.co.uk.

When I was a child, we kept chickens and ducks. It's great for kids to understand where their food comes from and, as chickens only cost 3p a day to keep and a pair can produce up to a dozen eggs a week, it is probably the most sustainable way for anyone to produce their own protein.

The Tibbetts were delighted with the new additions to their family. In fact, little Genevieve said to me, 'Do you know who my best friends are? The chickens. I love them already!' This was an hour after their arrival.

Not only were the kids delighted with 'Chicken Licken' and 'Henny Penny' but their new additions to the family could eat all their veggie waste.

When I was a kid we kept chickens and ducks. It's great for kids to understand where their food comes from.

Nappy rash

We Brits throw out around eight million disposable nappies every day, which then get taken to landfill sites. Not only is landfill space rapidly running out in this country, but many household items such as nappies and supermarket carrier bags could take up to 200 years to decompose. The most effective way of reducing our rubbish is to deal with it at source. That means buying a reusable shopping bag and using real nappies. That's not to say that real nappies are innocent of environmental crimes, but the landfill situation is an important one and disposable nappies are a substantial contributor.

Many people use disposable nappies because of the heavy-duty washing that comes with reusable terry nappies. But what about using a nappy-washing service? Instead of accumulating mountains of disposables that have to be buried in landfill, you simply hand over your own nappies, which belong only to your child. These are laundered for around £10 a week – about the same cost as buying disposable nappies – and are returned to you. Contact the National Association of Nappy Services (www.changenappy.co.uk). In addition, many local authorities now provide a one-off cash back to parents who opt for washable nappies. This payment is to help cover the cost of the initial outlay.

We Brits throw out around
eight million disposable
nappies every day.

A nappy-washing service
will cost you around
£10 a week
– the same cost
as buying disposable nappies

THE GREEN CLAIMS CODE

Every product we buy has some kind of impact on the environment – through its production, distribution, use and disposal. Some products are manufactured in such a way as to minimise their environmental impact, but confusing logos and misleading claims on packaging and in product literature can make it difficult to distinguish between products that are genuinely less harmful to the environment and those that make bogus 'green claims'.

What is a 'green' claim?

A green claim is any information provided about the environmental characteristics of a product. This applies to claims made in product literature, adverts, symbols, text or graphics.

The Green Claims Code is a voluntary scheme introduced by the government to provide guidelines for manufacturers when providing environmental information about a product. The Code aims to ensure that:

- Claims about the environmental aspects of a product are accurate and informative.
- Self-proclaimed green claims made by manufacturers are not misleading.
- Consumers are able to check the green claims made by manufacturers about their products.

FAIR AND ACCURATE GREEN CLAIMS:

- This paper comprises 80% post-consumer waste.
- Paper manufactured using 100% recycled post-consumer waste, collected from recycling banks.
- Pencils made from sustainable FSC (Forest Stewardship Council) accredited timber.

MISLEADING GREEN CLAIMS:

- 'Now contains twice the recycled content' – that's not so great if the product only contains 5% recycled material to start with. It is misleading to give the impression that a product as a whole is 'recycled' if that is not the case. If a product is 100% recycled, then it should be stated clearly and in plain language. If it's only 20% recycled, it's misleading to claim that the product is 'a recycled product'.

- 'Now uses less energy', without explaining why and how it uses less energy. It could be fair to say that the product 'uses 20% less electricity in normal use than our previous model', but any product comparisons need to be qualified.

- Beware of ambiguous claims such as 'sustainable', 'green' or 'non-polluting' and watch for vague descriptions such as 'friendly', 'kind' and 'safe' being linked with words such as 'earth', 'nature' and 'environment'.

BOGUS SYMBOLS

The use by a company of its own 'green' symbol or logo to give some kind of reassurance to consumers is potentially misleading unless it is accompanied by a clear statement, in line with the Code, that explains just what the image means. For instance, where the Möbius loop is used for claims of recycled content, the percentage of recycled content should be clearly stated. Symbols and pictures like trees, flowers, globes, the sun, etc. are big no-no's unless there is a direct link between the product, the object and the environmental benefit being claimed.

COMPLAINTS ABOUT GREEN CLAIMS

If you're baffled by green claims on a product, or you think the manufacturer is trying to pull the biodegradable nature-friendly eco-wool over your eyes, you can take the following action:

- Contact the head office of the manufacturer (or retailer if it's their own brand) and ask for an explanation of the claim and how it complies with the Green Claims Code.
- If you think the claim is simply untrue, or if you're not satisfied with the explanation given by the manufacturer or retailer, speak to your local authority trading standards department. Give them details of the claim and any other relevant information you have.
- Complaints about a claim on radio or TV adverts can be made to the Independent Television Commission or Radio Authority.
- If the material is printed or on the Internet contact the Advertising Standards Authority.

CHANGING YOUR SHOPPING HABITS

All of us like to have nice things, right? A growing number of us are using 'retail therapy' to counter our increasingly stressful and busy lives. Shopping can give us an instant 'high' and buying new things makes us feel good. However, for many, these spending habits are becoming increasingly difficult to manage.

Credit is much easier to come by and the advertisers are hell-bent on encouraging us to earn more and spend more. That's hardly surprising when you consider that advertising is essentially designed to create demand regardless of whether or not a product meets the genuine needs of consumers.

The typical 'shopaholic' is usually associated with the fairer sex, but a growing number of men are now spending compulsively. And whilst us girls tend to buy clothes, shoes, perfume and make-up, the guys are shelling out for all sorts of electronic equipment, sports gear and car accessories.

How can you tell if you're a 'shopaholic'? Well, if you're buying things you don't need, use or want or if you're spending more than you can afford, then you may have a problem. I'm not saying that we shouldn't ever go in a shop again, just because we're a little bit guilty of over-indulgence at times. However, 'shopaholics' are compulsive spenders who act this way most of the time. Shopping and spending become a preoccupation which eventually starts to disrupt their work, home and social life.

The only way to reduce waste and improve the environment is to start thinking about what we buy and where we buy it.

One of the easiest things we can do is to cut down on the number of trips we make to shopping centres – especially those that are out of town. By driving to these places, you're not only adding to traffic pollution and congestion, but the stuff you're buying isn't necessarily going to be cheaper. Just add up the amount of fuel you've spent getting there and paying for your parking – especially if you can't offset this against your shopping.

What can you do instead? Well, you could start by walking to your local high street to see what's on offer. Or if you don't fancy anything that energetic, why not see if you can get things delivered to your home instead. Remember, this is all about changing your habits and thinking in a different way.

ARE YOU ADDICTED TO SHOPPING?

Do you get an instant 'high' when you shop?	
Do you head straight for the shops when you're feeling wound-up or miserable?	
Do you buy things you already own?	
Do you buy things you don't need?	
Do you buy clothes or shoes without trying them on?	
Do you buy CDs that you never listen to?	
Do you have more than one credit card or store card?	
Do you hide things that you buy from your partner, friends or family?	
When you've really splashed out on something, do you lie about how much it costs?	
Do you often spend up to the limit on your credit cards?	

If you answer 'yes' to more than half of these questions, your spending may be getting out of control. Try working out what triggers your spending sprees. Do you head for the shops after a row with your partner or an upset at work? Have you seen a friend with some new clothes and feel that you need an instant image make-over? Are your shopping habits driven by feelings of low self-esteem? Once you're aware of why you feel the need to shop and spend, you'll be in a position to do something about it.

WHAT AND WHERE TO BUY . . .

If the thought of growing your own food all seems a bit too much like hard work, and you really can't face giving up the shops, then you need to know what and where to buy certain products to ensure you're 'doing your bit' for the environment.

SPECIAL DELIVERY

Because we're working longer hours and are leading increasingly busy lives, delivery services are becoming ever more popular. Why not try ordering a local organic vegetable and fruit box that comes with minimum packaging?

It's not just country folk who can get organic deliveries. If you're a busy Londoner, for instance, the Organic Delivery Company can bring all sorts of organic goodies straight to your front door. Delivery is free for orders of £13.95 and above and you can set up a regular order (either online or by phone) or just have a one-off delivery. Visit www.organicdelivery.co.uk to order.
Even the supermarkets are catching on to the benefits of delivery services. After all, one van delivering to a number of houses is better than hundreds of individuals jumping into their cars and heading for the stores.

There are lots of local delivery services and box schemes to choose from. Box schemes are a box of organic foods which come straight from the growers and are packed and delivered to your home – so, essentially there's no middlemen and no shops.

Visit the Soil Association's website www.soilassociation.org, which has a list of organic producers and suppliers.

SHOPPING ONLINE

Online shopping is a growing trend in this country. And not only does it save you time – and cuts down on traffic pollution through fewer car journeys to out-of-town shopping centres – but it allows you to find out more about a company and its products. You'll also have better access to product reviews and rating systems so that you can make comparisons between various products by reading comments from other purchasers. This, in turn, helps you to make more informed choices.

You can buy pretty much anything online, from food and drink to electrical goods and clothing. What's more, if you shop online you're not limited by global time differences, so you can shop when it's convenient for you to do so and not during the set hours that many shops have to adhere to. Other benefits are that you can customise your purchases and a lot of products are cheaper when you buy them online, as you're buying direct from the supplier. It's also good for elderly people or those who have trouble accessing shopping centres because transport isn't readily available.

FAIRTRADE

When you're next buying coffee, tea, chocolate and other imported goods, look for the Fairtrade logo. Fairtrade ensures that small farmers in developing countries receive a fair share of the money you pay for their product.

GROWING YOUR OWN FOOD

If you really want to avoid the supermarkets and shops, then why not grow your own food? Of course, growing your own food takes a bit of time, but what's more important than your health? Good wholesome, home-grown food is one way of ensuring that we're not filling up our bodies with the pre-processed muck that we so often grab in the shops because it's there, or because it'll save us time.

ALLOTMENTS

These are normally administered by your local council. You can rent a plot of land to grow all sorts of crops on and they are extremely cheap. For example, you can grow potatoes, corn on the cob, runner beans and all sorts of currants and berries, etc. It's a good way of getting to know other people in your local community, too. If you don't fancy doing all the hard work yourself, why not share it with friends and/or family. An allotment costs around £30 a year to rent. There is no packaging, no queues at the checkout and no unknown chemicals sprayed on your food.

ORGANIC

The Soil Association mark guarantees that the product is organic, and can be found on a whole host of goods. Organic farming does not use chemical fertilisers and pesticides, benefits the soil, the food, and the environment, so it's good to go organic if you can. It may cost a bit more, but often it tastes far better and it's also much better for your health.

MEAT

If you're a meat eater then try to buy organic or free range. These animals will have been reared under more natural conditions with less (or no) artificial hormones or pesticides. That means it's much better for you, for the environment and for the animal's welfare – at least when it was still alive! Organic meat can be expensive though, so why not eat a vegetarian meal a few nights a week and then splash out on an organic steak at the weekend?

FISH

The Marine Stewardship Council sets an internationally recognised standard to measure and reward sustainably managed fisheries.

At harvest time there is **nothing so rewarding** as sharing your glut of good food with friends, neighbours and family.

FARMERS' MARKETS

Farmers' markets have grown in popularity and number in recent years and there are now over 500 held around the country. All the food is locally produced, generally unpackaged and usually grown by the person selling it to you. Buying direct from the farmer who has grown the produce has many advantages, not least that you know where the food has come from. To find where your nearest one is visit www.farmersmarkets.net for a full list. If there isn't one near you there may be a farm shop. City dwellers don't have to miss out either. There are a number of city farms operating in a similar way; visit www.farmgarden.org.uk to find your nearest one.

WOOD

When you are buying wood products check for the Forest Stewardship Council (FSC) mark. If a product has that mark, it means that it uses sustainable timber from a well-managed forest. Alternatively, buy reclaimed timber which is fully seasoned and lacks the 'raw' look of new wood.

PAINTS

Many types of paints and solvents contain 'volatile organic compounds' (VOCs). These cause air pollution and are hazardous to your health, so buy low or VOC-free paints wherever possible. There are several suppliers of environmentally friendly paint, try The Green Shop at www.greenshop.co.uk. I have used a wide range of paints and finishes in my home. I have even used paint made from clay which, believe it or not, required only one coat, is quick to dry and comes in a lovely range of shades.

CLOTHES

Now I know that in the past 'vegetarian' shoes looked as if you were wearing loaves of bread on your feet, but things have come a long way since then! So, if you really want to wear your green credentials on your sleeve then look no further than www.greenfibres.com, which sells clothes made from organic fabrics. Try www.hug.co.uk for a range of cotton clothing. The baby wear is very cute – I've just bought some for my new grandson. And you can't get any more ethical than wearing something made from hemp. It grows without the need for pesticides and is biodegradable. Check out www.motherhemp.com for more details.

SEASONAL FOODS

Of course it would be lovely to eat strawberries all year round, but have you ever stopped to think about where the food you buy comes from? The chances are that banana you're about to munch has been flown in from the Canary Islands.

That's not good news for the environment. Why? Simple . . . there's a vast amount of fuel used in transportation, energy is wasted in its production and, therefore, large amounts of damaging carbon gases are released into the atmosphere. Who foots the bill? Would it surprise you if I said it was you? Somebody has to pay for that transportation, so that's another hidden cost that doesn't appear on your shopping bill.

Simply by stopping to think about where your food comes from, and making the right choices, we can make a world of difference.

Spring
- Asparagus
- Broccoli
- Brussels sprouts
- Carrots
- Leeks
- Lettuce
- Parsnips
- Radishes
- Spinach
- Spring cabbage
- Spring onions
- Strawberries
- Tomatoes
- Turnips
- Winter cabbage

Summer
- Asparagus
- Aubergine
- Beetroot
- Broad beans
- Carrots
- Cauliflower
- Courgettes
- Cucumber
- French beans
- Lettuce
- Mangetout
- Onions
- Peas
- Potatoes
- Radishes
- Runner beans
- Spinach
- Spring onions
- Summer cabbage
- Sweet corn
- Tomatoes
- Turnips

Autumn
- Apples
- Blackberries
- Blackcurrants
- Cherries
- Cooking apples
- Dessert apples
- Gooseberries
- Onions
- Peaches
- Pears
- Plums
- Potatoes

Winter
- Aubergine
- Beetroot
- Brussels sprouts
- Carrots
- Cauliflower
- Celery
- Cucumber
- French beans
- Leeks
- Lettuce
- Onions
- Parsnips
- Peas
- Potatoes
- Radishes
- Red cabbage
- Runner beans
- Swede
- Sweet corn
- Turnips
- Winter cabbage

PURCHASING POWER

If you really want to see purchasing power in action, then look no further than Ethical Consumer (www.ethicalconsumer.org). This organisation turns the spotlight on some of the major achievements brought about by consumer pressure over the years, and has information on all the latest consumer boycotts.

You may also be interested in *The Good Shopping Guide* published by Ethical Consumer. There are four main sections to the guide, covering the home, food and drink, money, and health and beauty, with further product subsections from tea to toasters.

Companies that produce the products featured in the guide are rated: good, questionable or bad, according to their environmental, animal-welfare and human-rights performance. Those that pass the test can add the Good Shopping logo to their packaging (e.g. Unite Energy Ltd).

If you would like to know more about the issues surrounding your purchases, you could also check out www.greenchoices.org, which has links to some of the best ethical companies and organisations on the web, selling everything from food to holidays.

For one-stop ethical shopping there's The Green Shop, which offers a mail-order service through www.greenshop.co.uk. It stocks a range of essentials of its own and many well-known ethical brands.

The most extensive one-stop ethical shop online is probably the Natural Collection catalogue at www.naturalcollection.com, a trading partner for Friends of the Earth.

The Centre for Alternative Technology's website www.cat.org.uk is perfect for those of you who wish to take your 'green' credentials one step further. There are courses and books on how to make your life more sustainable, and a range of energy-saving products such as radiator insulator panels and solar battery rechargers.

In today's world of disposable nappies, plastic bottles and PVC toys, alternatives for baby products can be hard to find. But, fear not, www.greenbabyco.com has all you could ever wish for.

Remember, beauty comes from within, so

don't make the planet suffer just for you to look good.

What you have saved . . .

WHAT YOU HAVE SAVED . . .	SO TREAT YOURSELF . . .
£430 of food a year from being a more conscious consumer and not throwing food that never gets eaten.	Buy luxury hampers for all your friends and family at Christmas.
The local shop – keep a local shop in business by buying fresher, organic goods with less packaging.	Save a piece of history and help keep the charm in your local town or village.
Expectations – you expect a product to last for life especially if you are paying good money, so stock up on cotton hankies, bulk buy eco-friendly cleaning products and long-lasting energy-saving light bulbs.	Expect more of yourself – the way you shop and every other way you affect the planet.

If you just do one thing . . .

There will always be someone who needs that treat more than you so, rather than making an impulse purchase for yourself, use your impulses to surprise someone with a gift.

CHAPTER FIVE

What's your poison?

The air quality in your home can be five times more polluted than the outside air.

As most of us tend to spend approximately 90% of our time indoors, good air quality in the home is vital for our health. It may shock you to learn that our homes are filled with a potentially poisonous cocktail of chemicals that can cause serious health problems.

We all need a constant supply of clean, fresh air to live and to maintain our health, but, because of our wasteful ways, we're polluting the air with car exhaust fumes, chemicals from factories producing our consumer goods and power stations generating our energy supply.

Air is considered to be polluted if it contains substances which could damage our health and the environment or which cause a nuisance. According to some experts, pollution can be just as bad indoors as it is outdoors. Yet, good air quality indoors is often overlooked as a means of improving our quality of life.

If anyone in your household regularly suffers from headaches, itchy or watering eyes, nose or throat infections, dizziness, nausea, colds, asthma or bronchitis, these symptoms could well indicate high levels of indoor air pollution.

Many of these symptoms follow a seasonal pattern, so continued monitoring of sufferers may show that symptoms are worse when the building is tightly sealed during winter, or when the weather is hot and humid. Symptoms may also occur when you're doing a major cleaning job. A sensible first step towards improving indoor air quality is to remove or reduce, where possible, as many sources of pollutants and toxins as we can.

Essentially, chemicals can be found in anything from soft furnishings, carpets and paints to household cleaning products, toiletries and cosmetics. As these toxins build up inside your home without you even realising they are there, you're probably wondering what you can do about them?

HOME: SAFE AND SOUND?

A certain amount of indoor air pollution is simply outdoor pollution that has found its way into our homes. But, normally, when we talk about pollution and toxins in our homes, the main problems are:

Damp, condensation and mould
Poor air quality
Allergens

IT'S GETTING DAMP IN HERE . . .

When we cook or take a bath or shower (and even when we're asleep) water vapour is produced. Condensation can occur through warm, wet air produced in kitchens or bathrooms being allowed to circulate around the house to unheated rooms such as bedrooms. In these conditions, mould may start to grow and when mould spores are inhaled, they can aggravate chest complaints and respiratory conditions such as asthma and bronchitis.

- Keep lids on saucepans when cooking to keep moisture under control.
- Keep kitchen and bathroom doors closed to prevent dampness spreading to other parts of the home.
- Fit mechanical extractor fans, preferably fitted with a humidistat, to kitchens or bathrooms where a lot of moist air is produced.
- Mould growth is a symptom of dampness, so to get rid of it you need to cure the dampness first – and if you can't work out what's causing dampness you need to speak to your local authority's Environmental Health Officer. Once you know what's causing the damp and once that's been remedied, you can clean and treat mould spots and redecorate, if necessary, using an anti-mould paint.
- Keep rooms well aired and try to avoid condensation by opening windows in bathrooms to let moist air escape after a bath.
- Ensure your home is properly insulated and heated; you may need a low level of heating even in unused rooms to stop condensation occurring.

PLANT PURIFIER

Plants are the oldest, most natural air filters on the planet. There are a number of common houseplants that have been shown to effectively detox households by absorbing poisonous vapours and releasing oxygen back into the air. A great source for helping to find plant purifiers is www.flowers.org.uk. Species to look out for include Mexican cacti, spider plants, ivy, rubber plants and gerbera daisies, which are all said to help humidify and freshen up stale environments.

IT'S GETTING STUFFY IN HERE . . .

There are believed to be up to 300 volatile organic compounds (VOCs) inside the typical household. These are generated from household cleaning products, paints, building materials, furnishings, carpets and tobacco smoke. VOCs are carcinogenic and long-term exposure to certain VOCs can lead to various cancers.

- MDF, veneered chipboard or plywood furniture emits toxic chemicals not usually present in solid wood furniture. Carpets and soft furnishings are also responsible for giving off chemicals such as formaldehyde.

- Another hidden danger in our homes is carbon monoxide – a colourless, odourless toxic gas that is produced by the incomplete combustion of fossil fuels such as natural gas, propane, heating oil, paraffin, coal, charcoal, petrol or wood. Any appliance that depends on burning these fuels for energy or heat such as gas fires, central-heating boilers, room heaters, water heaters, cookers or grills can produce carbon monoxide if they are not installed, maintained or used correctly. Because this gas has no warning properties it is often called 'the silent killer'. Initial symptoms of exposure may include dizziness, excessive tiredness and fatigue, headaches, nausea and irregular breathing. However, death from poisoning can occur without any of these symptoms being experienced – a person may simply fall asleep and fail to regain consciousness.

- Formaldehyde isn't just something that trendy artists use to preserve dead sheep in; it's a VOC that can be found in any number of household items, including foam insulation, chipboard furniture and emulsion paints. At high concentrations, formaldehyde can cause eye and airway irritation, neurological damage, asthma and cancer.

CHECK YOUR BOILER

Carbon monoxide poisoning from faulty boilers and gas fires affects hundreds of people each year and can be fatal. Sooty marks on the wall and yellow flames instead of the usual blue ones are all signs that your appliances could be faulty. If this sounds familiar then you should have your heating checked out by a CORGI-registered engineer as soon as possible and ask your doctor for a diagnostic blood test.

IT'S GETTING CREEPY IN HERE . . .

Allergens such as house dust mites are tiny, microscopic creatures that live in carpets, bedding and soft furnishings and are believed to be one of the major triggers for asthma and eczema. Mites like warm, damp conditions, so regular vacuum cleaning and high-temperature washing of clothes when necessary can help control their numbers. Don't be tempted to run high-temperature washes just for the sake of it, though.

PERFECT PETS

Despite Britain being a nation of animal lovers, loads of people are allergic to cats and dogs, as their hair and fur can trigger allergies in susceptible people and can make existing conditions such as asthma or hay fever worse. To keep pet allergens at a minimum, make certain rooms pet-free zones and remember to wash and groom pets regularly. The Cat Protection League and Canine Defence League can provide you with tips on how to make life easier for anyone with allergies or asthma exacerbated by pets. Or contact the Asthma UK adviceline 08457 010 203.

DUST MITES

Dust mite excrement can build up in mattresses, duvets and pillows: 70% of the weight of your pillow is likely to be dust mite droppings according to some experts. Use a steam cleaner on mattresses to kill bugs without using chemicals.

Your house dust is made up of dirt pet hair, dust mite excrements and human hair and skin.

SOLVENT ABUSE CAN KILL INSTANTLY

The use of this statement is voluntary, so you won't find it on every product. It normally appears as a separate warning on many solvent-containing products such as spray adhesive.

CHILD RESISTANT

Products must have child-resistant closures (lids, caps, etc.) on any products labelled 'very toxic', 'toxic' or 'corrosive'. There is also a list of specific chemicals that, if used as an ingredient, automatically means a manufacturer must use child-resistant packaging. However, these closures don't necessarily mean that the product is child-proof – so keep all such products out of the reach of children.

INGREDIENTS

Since 1989 there has been an EU recommendation on ingredient labelling for detergent and cleaning products. Certain ingredients should, therefore, be labelled if they're present in the product above a concentration of 0.2%. Enzymes, preservatives and optical brighteners – found in some washing powders – should be labelled regardless of the concentrations.

MANUFACTURER'S DETAILS

The full name, address and telephone number of the manufacturer must be printed on all dangerous products.

ELBOW GREASE

According to a recent survey, it is women who spend an average of 21 hours a week on household chores, which suggests that women make most of the decisions when it comes to what sort of household cleaning products we buy. Organisations such as the Women's Environmental Network are encouraging women to minimise waste and to use more environmentally friendly products. Many cost less than a fiver. Visit The Green Shop at www.greenshop.co.uk and buy Ecover products at www.ecover.com. It is also worth visiting Earth Friendly Products. (www.greenbrands.co.uk)

Household borax

Borax is very effective as an antibacterial, cleaning, fungicidal and bleaching agent. If you replace your usual household cleaning products with borax it's better for your health and the environment. You can buy borax from The Green Shop and it can be used for a number of household cleaning jobs – here are just a few:

- Add half a cup of borax to your washing with the usual amount of washing powder to boost its cleaning power and deodorise the wash.
- Soak delicates in a solution of one or two tablespoons of washing powder and a quarter-cup of borax in a bowl of warm water. Rinse in cool water and dry.
- Dissolve one tablespoon of borax in a litre of warm water and use the solution to wipe the fridge clean and deodorise it at the same time.
- For wine and other liquid stains on carpets, dissolve half a cup of borax in half a litre of warm water, leave for half an hour and sponge off. For odours dampen the area, then sprinkle with borax and hoover when it's dried.
- Borax can be purchased from www.dripak.co.uk, Wilkinsons or Robert Dyas. They also stock soda crystals which can be used for a multitude of cleaning tasks.

dry place.
This product
been tested
This box is
minimum of 90%
and is 100%

THE USUAL SUSPECTS

A recent study found that each of us is typically walking around with more than 300 man-made compounds in our bodies. Nobody knows what long-term effects these chemicals will have on our health so it pays to know what the main culprits are and how we're exposed to them. Once we know what and where, we can minimise our use of them.

PHTHALATES

These are a group of chemicals strongly suspected of being hormone disruptors. While they've been banned from baby toys in seven European countries, they're still used in the UK. The only legal restriction on putting phthalates in toys is a rolling six-month EU ban on their use in toys 'intended to be put into the mouths of under-threes'. Some baby toiletries also contain phthalates.

BISPHENOL A

This is another endocrine-disrupting chemical found in polycarbonate plastics, which are used to make baby feeding bottles, water bottles and food storage containers, and the lining of food tins. Most baby feeding bottles sold in the UK are made from polycarbonate plastic derived from Bisphenol A, but manufacturers say that polycarbonate bottles are safer to use than glass feeding bottles, which can break.

ARTIFICIAL MUSKS

These are used in all kinds of fragrance products, cosmetics, air fresheners and shower gels. For instance, many air fresheners contain hormone-disrupting artificial musks (sometimes loosely referred to as 'parfum' on the label). You'll also find artificial musks in some baby creams, wipes and other cosmetic products and toiletries.

ALKYPHENOLS

These are used in detergents, paints, glues and lubricating oils as well as in some cosmetics from shampoos to shaving foams, and they become more endocrine disrupting in the environment when they break down. These breakdown products have been linked to human fertility problems.

ORGANOTINS

These are also known as alkyltins and are used in about 8% of PVC products produced in Europe. PVC products containing endocrine-disrupting organotins include carpet lining and vinyl flooring. They're also used as fungicides in replica football shirts and the insoles of trainers. The use of organotins in disposable nappies (or any PVC products) isn't regulated.

Fortunately there are some responsible manufacturers out there, so you do have a choice. To help you, Friends of the Earth provides a list of all the major high-street retailers who were prepared to respond to a questionnaire on the safety of their products. Visit www.foe.co.uk to see the full results.

Also, visit Greenpeace's Chemical Kitchen at www.greenpeace.org.uk, which reveals well-known brands and suggests safer alternatives where they can.

HOW WE USED TO CLEAN

If you don't want to spend money replacing your household cleaning products with eco-friendly ones, then why not try some timeless, natural alternatives to the usual chemical cocktails? They'll save you money as well. Here's what and how you can use what's already in your kitchen cupboard:

KITCHEN BINS

Kitchen bins should be emptied as soon as their contents start to smell and then should be washed out once a week. You could try using a solution of one teaspoon of borax in 500 ml of hot water as a disinfectant, rather than swilling it round with bleach. Alternatively, thyme oil and salt both have disinfectant properties.

KITCHEN CUPBOARDS

Kitchen cupboards should be cleaned out regularly several times a year. Take out all the food from each cupboard and throw away any items that are past their sell-by date. Then, wash the interior of the cupboards with a solution of bicarbonate of soda and warm water and wipe dry with an old towel. Leave the cupboard for a couple of hours to dry thoroughly before restocking with food.

OVENS

A paste of baking soda and water left on for five minutes and then washed off with a scouring cloth and hot water is an effective oven cleaner. Or you could sprinkle salt on spills while they are still warm to ease cleaning once the oven is cool.

FREEZERS

The more you open the door of your freezer, the more you'll need to defrost it. As a general rule, you should defrost your freezer when the ice reaches a thickness of 5 mm. When defrosting is finished, dry the interior then rinse with a solution of bicarbonate of soda (15 ml of soda to 1 litre of water), and dry with a clean towel. If smells linger, fill the freezer with crumpled newspapers and leave it switched off with the door slightly open for a couple of days. Newspaper helps to absorb smells.

FRIDGES

If your fridge develops a pong because of rotten food, or if you've forgotten about that smelly cheese you bought on your day trip to France, clear out all the food and wash the interior several times with the same bicarbonate of soda solution mentioned above. You can wash shelves and other bits and bobs in warm water and a solution of borax.

Once or twice a year, make sure you pull the fridge away from the wall and dust the coils at the back to remove dust build-up, as this will make your fridge less energy efficient and will add pounds to your electricity bills.

KETTLES

Furry deposits can build up on kettle elements – especially in hard-water areas. It's best to descale kettles before the fur has had a chance to build up. Distilled white vinegar is good for removing scaly deposits in kettles (as well as lime scale in toilets or sinks). And if you use a teapot, then lemon juice is good for removing tea stains.

BATHROOMS

Lime-scale build-up and blue-green marks on enamel baths are a sure sign that you've got a constantly dripping tap. So, first things first, you need to get that fixed, as you're wasting water. To get rid of stubborn stains, use bicarbonate of soda to scour sinks and baths and to treat tide marks. Mouldy areas that appear around bath sealant can be treated with a solution of borax and then scrubbed with an old toothbrush.

DRAIN DAMAGE

Dissolve a quarter of a cup of baking soda and 50 ml of vinegar in boiling water and use to unblock drains.

TILES

Wipe over with a white-vinegar solution (one part vinegar to four parts water), then rinse and wipe dry.

LOOS

Daily brushing and flushing and a once-a-week clean with borax should be sufficient for loos. Bleach damages the ceramic glaze of the pan and makes the loo more difficult to clean. Hard-water build-up should be covered with a thick paste of borax and vinegar, left for a couple of hours, then brushed off and rinsed.

SHOWERS

If shower curtains are prone to mildew, soak them in vinegar and water and then rinse and machine wash (check the label first). Hard-water deposits on shower walls can be treated with neat white vinegar – leave for 15 minutes and rinse off. If your shower head is clogged, soak all the pieces (apart from the rubber washer) in neat white malt vinegar. Use an old toothbrush to brush away sediment build-up and chalky deposits before putting it back together again.

WINDOWS

Try using two tablespoons of distilled white vinegar and some borax in a spray bottle for cleaning windows. Some smearing may occur on first use due to waxy build-up from previous spray cleaners, but the borax will help to remove this.

LAUNDRY

If your whites have got very grubby, try soaking them in a 5% dilution of lemon juice and water and leave outside in the sunlight. Then, add 30 ml of bicarbonate of soda to the washing water and a squirt of lemon juice in the rinse for glorious whites. If black clothes have faded to murky grey, soak them in warm water with a little vinegar to restore them to their former Goth glory!

If the clothes you've been wearing during the day are not dirty, they could do with an airing before you put them away as all clothes tend to pick up smells. Hang your clothes up by the shower or preferably outside for a couple of hours to refresh them – your clothes will last much longer, too, as they won't be constantly knocked about in the washing machine.

SODA
CRYSTALS
SODA
CRYSTALS

CASE STUDY

LAMBERT FAMILY
Peter, wife Karen, mum-in-law Pam, and daughters Jade (19) and Krystle (21)

When I visited the Lamberts I must admit to feeling a little bit sorry for Peter. He's the only man of the house with four women who drive him mad with their constant use of hair and beauty gizmos and gadgets – which they always forget to turn off. Their electricity bill was £65 a month – nearly three times more than the national average.

They were certainly a clean family – racking up a long list of unnecessary consumer products to maintain their pristine lifestyle. Mum Karen and mum-in-law Pam were perhaps a little over-obsessed with the house being immaculate and smelling nice. Homemade pot pouri replaced the plug-in air fresheners they had in every room. They used the vacuum cleaner for 30 mins a day, and the washing machine for 18 hours a week on average. They also liked the house to be unbearably hot – the heating was on 24/7 (partly because Pete didn't know how to use the timer).

Krystle wouldn't answer the door without her make-up on. She loves wearing the latest fashions, and won't be seen dead in the same outfit twice. Rather than recycle her clothes, she'd bin them and buy new ones. As a member of the cabin crew of a budget airline, she's also clocking up the air miles – polluting the air with even more fumes! And Peter's prized possession is his Jacuzzi which he was heating to 39 degrees, for all hours of the day – ready for him to jump into every morning for his hour-long soak.

The whole family had to cut down on household products and I thought the best way to start was changing the way they clean. I set Karen and Pam to work and asked them to de-clutter their kitchen cupboards, which were stuffed with harmful cleaning products. I made them use old-fashioned remedies for cleaning instead such as vinegar, newspaper, salt, lemons and bicarbonate of soda – all of which Pam had used years ago. They were delighted with the results and eventually weaned themselves off all of their toxic products.

DISPOSING OF DIFFICULT AND HAZARDOUS HOUSEHOLD WASTE

My golden rules up to now have been to reduce, reuse and recycle wherever possible, but there are a number of household hazards that need to be disposed of carefully. Up to 5% of household waste can be classed as hazardous due to its potential to harm human health and the environment.

The following should help you to identify what are the major hazardous materials and how to get rid of them safely.

POLICE YOUR POISONS

- Minimise the amount of hazardous waste you generate by buying materials you know are less hazardous or by using alternatives. For example, paints, varnishes, glues and cleaning and treatment agents that are plant-based, water-based or low in solvents are good alternatives.
- Buy more durable products such as rechargeable batteries – don't go for disposables where you can help it.
- Purchase only what you need – including the right quantities of cleaning products, detergents, etc. or source more eco-friendly alternatives.
- Always follow the usage and disposal instructions.
- Donate unwanted or broken furniture and fridges or freezers to charities and community organisations that repair and redistribute them.

GARDEN AND HOUSEHOLD CHEMICALS

- There is a wide range of garden chemicals that can harm the environment if they're not disposed of correctly. These include pesticides, herbicides and fertilisers. These should not be poured into drains, sinks or toilets as this causes water pollution, particularly if large quantities are disposed of.

- Some pesticides are also toxic to pets and wildlife and will contain warnings on the packaging, including recommendations for how to dispose of them safely. If there are no guidelines for disposal, you should contact the manufacturer directly for further advice.

- Garden chemicals should always be stored in their original packaging so that their usage instructions are always to hand. Existing quantities should be used up completely wherever possible. Because of the chemical residue they contain, empty garden chemical containers can't be recycled.

BATTERIES

Most batteries contain heavy metals and only a very small percentage of consumer-disposable batteries are recycled. Most end up in landfill sites. But, some local authorities collect household batteries as part of their multi-material kerbside collection schemes, so it pays to check first with the local authority recycling officer to see what's available in your area.

In addition, some retailers and DIY shops have battery collection points near their stores. Rechargeable batteries can also be recycled once they have reached the end of their useful lives. Contact REBAT (www.rebat.com) who manage and collect most types of portable rechargeable batteries in the UK.

MEDICINES

If you have a good old rummage in your bathroom cabinets and drawers, then no doubt you'll find stockpiles of medicines and pharmaceutical products that have passed their use-by dates or are no longer needed. Whatever you do, don't chuck them out with your normal rubbish or pour them down the sink or toilet. All unwanted medicines should be kept in their original packaging and taken back to the chemists, who will dispose of them safely for you.

ASBESTOS

In the past asbestos was used in all kinds of household materials such as insulation, roofing and in appliances such as cookers and electric blankets. It's not used as frequently in modern buildings, so it's more likely to be found in older buildings or in outbuildings such as garages and garden sheds.

Your local authority waste disposal officer will provide advice on the handling and safe disposal of asbestos, as local authorities have a legal duty to provide facilities for household asbestos disposal.

FRIDGES AND FREEZERS

If your fridge is in working order, you can donate it to a community or charity organisation, many of which will collect it free for redistribution. Contact the Furniture Reuse Network (www.frn.org.uk) fridge collection service, which guarantees that up to 15% of the units donated will be reused and passed on to low-income families. If you live in London or Merseyside you can also contact CREATE UK for collection and recycling/reuse of refrigeration equipment. (info@create.org.uk)

Local authorities are obliged by law to provide facilities for householders to dispose of unwanted fridges and freezers. Usually, local authorities provide a collection service for bulky items (there may be a small charge for this). However, items collected this way will normally be sent to a landfill site. Alternatively, householders can take unwanted equipment to their local civic amenity and recycling site. Contact your local authority for further details.

OILS

For all you motorists who do your own oil changes, your unwanted engine oil needs to be disposed of very carefully. In fact, it's illegal to pour it down the drain and if you're caught you could be given a hefty fine. That's not surprising when you consider that the oil from one car oil change can pollute the surface of a lake the size of two football pitches. Unwanted oil can be recycled and purified for reuse, so you can take unwanted engine oil to a local garage which operates a recycling scheme or contact your local authority recycling officer for further advice.

When disposing of cooking oil, it should not be poured down the sink as it can block the drains and attracts vermin such as rats. Contact the Environment Agency for further information. You could save up the waste fat to mix with seeds, nuts and raisins to make a 'bird cake' to help feed the birds in the winter. Fat traps can be purchased from www.lessmers.co.uk. Remember that small quantities of cooking oil can be composted – but make sure it's cold before you compost it.

The Environment Agency runs an Oil Care Campaign, which will give you information on the nearest recycling facility or garage recycling scheme. Visit their website at www.oilbankline.org.uk or contact them on 0800 663366.

PAINT

Although not strictly hazardous waste, household paints contain VOCs, which can be harmful to human health if substantial amounts are inhaled. Again, paints or paint thinners should not be poured down drains. If you have small quantities of residual paint then it's best if you just allow it to dry out naturally or use old newspaper to soak up any excess. Containers can then be disposed of at your local recycling centre – but check with your local authority first, as these facilities may not be available at your standard drop-off point. If you've got larger quantities that you no longer need, then you can donate them to a Community Repaint scheme (a network of paint reuse schemes operating throughout the UK). The organisation offers a free collection service and distributes reusable paints free of charge to community sector groups and public agencies. SWAP (Save Waste & Prosper Ltd) co-ordinate the network and they can be contacted on 0113 243 8777 or visit their website at www.swap-web.co.uk for more details.

Poisoning: What to Do in an Emergency

It goes without saying that you need to make sure that all medicines, cleaning products, garden chemicals, etc. are stored in a safe place where kids can't get at them. It's good practice to regularly review your storage arrangements to make sure that harmful substances are out of reach of children.

Heaven forbid that anyone in your house should actually ingest any of this muck – but unfortunately accidents do happen. If someone in your house does swallow something poisonous you will need to do the following:

- Give them frequent sips of water if they have swallowed a corrosive liquid such as bleach.

- If they're unconscious, check that the airway is clear of vomit and tilt the head backwards. If they're breathing, put them in the recovery position (see box).

- Do not induce vomiting – if it burns going down, it will burn coming back up and could do more damage to airways.

- For an inhaled poison, take the person into the open air.

THE RECOVERY POSITION

- Turn casualty onto their side.
- Lift chin forwards in open airway position and adjust hand under the cheek as necessary.
- Check casualty cannot roll forwards or backwards.
- Monitor breathing and pulse continuously.
- If injuries allow, turn the casualty to the other side after 30 minutes.

For a baby less than a year old, a modified recovery position must be adopted:

- Cradle the infant in your arms, with their head tilted downwards to prevent them from choking on their tongue or inhaling vomit.
- Monitor and record vital signs – level of response, pulse and breathing – until medical help arrives.

BEAUTY SCHOOL DROP-OUT

Would you rub chemicals into your wrinkles?

Would you brush your teeth with washing liquid?

Would you dye your hair with petroleum?

Remember, every product you put on your body is made up of ingredients and many of these are believed to be endocrine disrupters found in household cleaning products and other toxic products. Daily application means that these chemicals can bioaccumulate in our body over time. Scientists are still unsure of the exact health implications, but safe to say, we could all do with cutting down the amount of synthetic chemicals we douse are bodies in.

Open your bathroom cupboards, look on your dressing tables and empty your gym bags. I wouldn't be surprised if you counted twenty different personal products that you use on a daily basis. In the UK we spend nearly six billion pounds on personal care every year and we are flushing these chemicals down the drain and into the sea.

For example, many people do not realise that some toothpastes containing flouride should be treated as a medicine and kept out of the way of children. Read the back of the toothpaste packet – it will advise you to keep it out of the way of kids, that kids should only use a pea-sized amount and that they should be supervised using it. Fluoride is actually a poison so try to buy toothpastes which use less harmful products. Any dentist will tell you the best way to keep your teeth healthy is with regular brushing and flossing and to avoid sugary foods and drinks. Weleda stocks non-fluoride toothpaste and with a little careful shopping you can find other brands on supermarkets and chemists' shelves. For further info on Weleda toothpaste and products try www.weleda.co.uk.

The best advice for buying beauty products packed with chemicals is to strip down every claim of the advertising. Do you really need the products you buy? Won't a natural ingredient do just as good a job? Hypoallergenic doesn't guarantee you won't have an allergic reaction – common sense tells us that even natural ingredients could cause an allergy. The marketing of toiletries has no regulations to say how 'organic' or 'natural' a product is. Are you being conned?

THERE'S SOMETHING IN THE WATER

Pure water does not contain any toxins, but over the past fifty years our drinking water has had more and more additives incorporated into the mains supply; these may spoil the taste, even though they rarely cause anything more than minor health problems. Drink bottled water if you must and improve your tap water in the following ways:

- Find out exactly what is in your water from your water supplier (listed under Water in Yellow Pages).
- Avoid drinking from the hot tap. Hot water is usually stored in a plastic or metal tank, and as water is a natural corrosive the surrounding material may well have seeped into it.
- A less expensive option is to filter water. Special active-carbon filters, which work by removing chemicals as they pass through them, can be plumbed in under the kitchen sink or attached to the end of a kitchen tap. Alternatively, use a jug filter – a plastic lid containing an active-carbon filter fits on top of the jug and filters the water as it is poured through.

In the UK we spend nearly six billion pounds on personal care every year and we are flushing these chemicals down the drain and into the sea.

What you have saved . . .

WHAT YOU HAVE SAVED . . .		SO TREAT YOURSELF . . .
You could save up to £30 a month (£360 a year) by using natural cleaning and beauty products.	→	Have fun making your own beauty products and have a pampering session wth your girlie mates.
Your child's allergy problems.	→	Avoid endless waits in the doctor's surgery. Take advantage of their new allergy tolerance and enjoy a day outside.
Your pride – be truly happy with your home and yourself knowing that you have achieved real results with simple, natural products.	→	Hold your head up high and look further afield. Feel confident that you and your family will be healthy enough to enjoy the planet 20 years from now.

If you just do one thing . . .

Make a habit of reading products. If it says 'harmful to your health'; don't buy it.

CHAPTER SIX

Less haste, less waste

Being wasteful in the home isn't just about plastic packaging and leaving the lights on all day; it's also about how we waste time, physical and mental energy, and personal space.

If you banish clutter, you'll feel more relaxed, less stressed and you'll reclaim your life!

TIME TO SPARE

Just stop for a second and take a good long look around your home. What do you see? Is it your inner sanctum of calm? Is it somewhere you can relax and unwind after a long hard day? Are you surrounded by the things you love: comfortable furnishings, personal trinkets and things that give you pleasure? Or, are you steadily drowning in clutter?

- We need to guard our time carefully, otherwise our 'free time' can suddenly vanish because we find that we have to go and visit the in-laws, or take those forms we meant to post to the local tax office. The secret is that you need to learn to plan ahead and make sure that everyone knows what you're planning – otherwise you'll find that there's an ever-increasing list of demands being made on your time and you end up having no time to do what you want to do.

- Be selfish with your time. Make sure that you block out a few hours for spending time with family or friends. That way, you're much more likely to stick to it. It doesn't matter if you haven't made any definite plans like going to play football in the park with the kids. If it rains, you can rent a film instead and sit and watch it with the family. The point is, you've allocated that time for you and for them and you're sticking to it.

- ME, ME, ME. Above all, being organised and taking control of your life will maximise the time you can spend enjoying life, so make sure that you take five or ten minutes every day to plan what you want to do with your 'ME' time and make a list. This will help to keep you focused and you're more likely to stick to it.

- No interruptions. Turn off the mobile phone and leave the answer machine on at home when you've set aside some time. If the call's not urgent, it can wait.

- How often have you thought to yourself, 'if only I had more space?' or how many times have you turned down a night out with friends because you 'don't seem to find the time, these days'? Most of us would like to spend more time with our friends and families and do the things that we want to do, when we want to do them. But, something called 'life' seems to keep on getting in the way.

- No doubt if you've got kids you'll spend a great deal of your time cleaning, washing and generally running around picking up after them. With toddlers there's all the toys and clutter they create to deal with, too. No wonder you feel frazzled, but if you put all their stuff into a large storage box at the end of each day, regardless of the fact that it'll be strewn all over the place by nine o'clock the next morning, you'll feel better for it.

- It's not about squeezing more and more into your already hectic schedule. It's about planning how you would like to spend your time so that you can improve the quality of the time available to you.

DECLUTTER DETOX

We're all leading increasingly busy lives and that means spending less time doing the things we want to do – and having less time for the ones we love. Decluttering your life doesn't have to take time and every minute you put towards these life changes will be pennies in your pocket, less stress and more time to do what you want to do.

WASTE OF SPACE

We've all got clutter – it's the sort of stuff we poke away in drawers telling ourselves that 'it may come in useful, one day'; or it's that old scarf you used to wear at university in 1971 regardless of the fact that you're now 57 and wouldn't be seen dead in it. It's the shoebox full of photos of people you met on holiday but never hear from and can't actually remember their names. It's the 24 second-class return tickets to Hove that you kept when you were temping, because you meant to get your travel money paid for, but forgot to send in your claim form.

All this stuff is clutter and it's wasting space. It's wasting your time, as you'll forever be rummaging through it all to find things that you really need and have forgotten where you've put them. So, it's time to cut it loose.

MAKE A CLEAN SWEEP OF THINGS

If you don't know where to begin, it may be time for what I call the 'home detox'. Just like our bodies, sometimes our homes need a complete clear-out before we start putting good theories into practice and operating more efficiently.

Look at your house – you are the best judge. If you think it's worth giving your home a complete overhaul then try these spring-clean tips to give your house a break from the wear and tear of day-to-day life. Once you've done this, your home will be ready for any waste-saving initiative.

- Let in some fresh air! Closed windows provide insulation and stop us wasting heat – but it shouldn't mean that you should live in a hermetically sealed box. Household pests such as dust mites thrive on warm humid atmospheres, so fling open those windows.

- Start with a general whip round to chuck out any dead flowers, empty ashtrays, put dirty cups in the sink and so on. Then vacuum or sweep your floors first before tackling the dusting.

- Tackle one room at a time and don't move to another one until you're done with the first one. You'll feel much more positive if you can complete one whole room, whereas you'll simply lose the will to live if you've not managed to wash the floors in all ten rooms having set aside the whole afternoon.

SETTING YOURSELF A CLEAN ROUTINE

It's easy to see what the major cleaning jobs are in our homes, so it's best to leave those for when you have time to do them. Having said that, don't let things build up so that when you get around to cleaning them they're so grubby it takes a day and a half to get them back to their former glory. The best way to develop a cleaning routine is to divide your home and its contents up as follows:

- Things that need a daily clean
- Things that need a regular clean
- Things that need an occasional clean
- Stuff that needs to be blitzed once a year

Most of us would like to spend more time with
our friends and families,
but something called 'life'
seems to keep on getting in the way.

TOP TIPS TO DECLUTTER YOUR HOME

START SMALL:
If the thought of decluttering your home sends you running for the hills, start by tackling small tasks first. Try setting yourself a ten-minute goal and sort out your underwear drawer. Tip it all out onto the bed; take a good look at each item and ask yourself – do I really want to keep this? If underwear hasn't been worn then pass it on to a charity shop. If you've got three pairs of Bridget Jones-style grey belly-huggers, get rid of them!

KEEP IT UP:
You've started now, so you might as well carry on! If you need to lift your spirits to face the next task, put on your favourite CD and get stuck in. If you're in an upbeat mood and are feeling good, you're more likely to carry on. Don't get too overambitious though and vow to sort out the entire loft, upstairs bedroom and the spare room in one morning. Set aside 20 or 30 minutes a day for a task and make sure you stick to it.

DON'T GO IT ALONE:
Get the rest of the family or friends involved. If you absolutely detest that ra-ra skirt you bought when you wanted to look like the lead singer of Bananarama, your friend (or daughter!) may love it. Unwanted items can be passed on to charity shops, friends, family or recycled. Or you could make yourself a few quid by getting rid of stuff at a car-boot sale. And don't forget eBay and *Loot*!

THE BANISHMENT BLUES:
It's perfectly normal to feel guilty, or even a bit sad, when you're clearing out personal belongings. You may feel guilty that you're giving away the hand-knitted shawl your aunty made for you last Christmas, even though you detest it. Or it may feel wrong, somehow, to get rid of old photos of drunken people whose names you can't remember. Don't be a slave to sentiment. It's OK to be sentimental, but you must be selective too. For the things you really cannot bear to part with, why not put them in a special keepsakes' box. Or pass on a cherished item to someone who you know will treasure it and give it a good home.

TRY THE THREE-BAG SYSTEM:
In bag one, put all the stuff you want to keep but won't use for a while. That should go into storage. You can then review this in six months' time; if it's still not being used, ask yourself why you're hanging on to it and if somebody else would make better use of it instead. In bag two, put all the things you want to repair, reuse or recycle and make sure you get on and do it. And in bag three, put all the stuff you plan to take to the charity shop.

REFUSE:

There's no polite way to refuse an unwanted gift, but if you really can't stand that china owl ornament with its malevolent red eyes and want to smash it every time you look at it, get rid of it. It'll only drag your spirits down and make you feel miserable if you keep it around.

STORAGE SOLUTIONS:

Invest in storage or, even better, keep cardboard boxes and create your own. Make sure you store like with like and can access what you need and want to use most.

CLUTTER-FREE KARMA:

By getting rid of what you don't want or need, you'll enjoy the benefits of a more relaxing, stress-free home. You'll be able to find what you need when you need it. You won't need to rush around frantically shovelling things in drawers to hide the mess from that unexpected visitor. And you'll be making the best use of space in your home, too.

CREATIVE DEJUNKERS

Don't dither – deal with it. Did you know that most bits of paper get picked up 10–20 times before they're dealt with?

Make a giant collage with your best and funniest photos or old postcards; paste them to a board and put them in a large clip-frame. That way you can enjoy them every day and make them a feature of your home.

If it's broken or you simply can't stand to look at it, get rid of it. You should only have things around you that you love and that make you feel good.

Keep bulky items like coats, shoes and bags by the front door so that when you're all tearing about in the mornings and are rushing to leave the house you can just grab stuff and go.

Don't put it off any longer;
confront your clutter today

CUNNINGHAM-LOCKE
Dad Jeff, mum Caroline and their children Chloe aged 7, Rhys aged 6 and foster daughter Karen aged 12.

When I arrived, I was greeted with a back garden that was strewn with building gear, toys and general rubbish. The loft had to be cleared to prepare it for insulation, but it was crammed with stuff they hadn't even looked at in years and that meant they had to get to grips with some serious decluttering.

Rather than sending it all to landfill, I arranged for a local recycling company (Ivor Skips 020 8575 6861) to drop off a skip. The family put all the stuff they no longer needed – which included toys, electrical equipment, wood and clothing – into the skip. There was a cost of £180, but the recycling company then sorted out the various items and sent them for reuse, repair or, if the item was beyond repair, for appropriate disposal.

With a clear loft and garden, they were able to make use of valuable space that had been lost to them. The properly insulated loft space meant that the kids could have a new play space and that the parents could enjoy more peace and quiet in the knowledge they had dealt with unfinished business by decluttering. They also painted their lounge with an eco-friendly paint and managed to finish jobs they had been putting off for ages.

It was not the calmest of homes – dad was addicted to his stereo, Rhys to his computer games and the girls to the TV. With so much noise and distraction there was a lot of shouting mostly because they couldn't hear what each other were saying. I imposed a ration on the amount of time the TV, games and stereo could be on. They were horrified – what would they do with their time? I suggested they do more as a family: play games, get some jobs out of the way and . . . talk! On my return, just one week later, it was definitely a more relaxing place to be. They had worked as a team and found that talking and listening without the distraction of all that white noise was interesting and pleasurable.

Decluttering really does add to the quality of our day-to-day life. A solid weekend of work, cracking through that list of jobs brings something priceless – a sense of calm and achievement.

TIME FOR THE IMPORTANT STUFF

If you are going to make the lifestyle changes outlined in this book then you need to change the way you think about things generally. Caring about the environment doesn't just have to be a 'do-gooders' ethos; it can also highlight the smaller things that really matter to you – things for your own selfish needs.

Choices, choices, choices

We all make choices. For example, if you work late you're not going to be able to go out for dinner with your friends. Sometimes that's just unavoidable. But, if it's become habitual, then you really do need to stop and think about your main concerns and managing your time.

The best way to cope with all the competing priorities we have to deal with on a daily basis is to get some sort of balance in life. You need to take stock; evaluate what's important in your life and then make some hard choices. Rather like decluttering our homes, making 'life choices' isn't easy. It often involves letting go of things (or people). There are no right or wrong choices, either – but sometimes we need to be a bit selfish and that can make us feel guilty. However, everything has its price and only you can decide what's worth paying for and whether or not the price is right for you.

You need to take stock; evaluate what's important in your life and then make some hard choices.

WHAT ABOUT SPONTANEITY?

For some of you, the idea of marking out your life like this will sap your will to live! 'What about spontaneity?' I hear you cry. Well, planning your time doesn't necessarily mean you can't be spontaneous. However, it's the spontaneous acts that cause us to make lazy choices around the home – 'I'll just throw this wine bottle in the bin, no time to recycle, I'm going to open another!' If we keep acting spontaneously without planning our future, we're going to ruin the planet. The trick to embracing plans is to make sure that you block out a set amount of time and then use it in a creative way. If you decide you'd fancy a night out, make sure you have the time available and do what you like at the last minute.

Rituals and routines

The faster our pace of life becomes, the less time we seem to be spending with our friends and families. Reintroducing some of life's little rituals back into your routine can help improve your personal relationships and lead to happier and more fulfilling lives.

For instance, when was the last time you and your partner enjoyed a meal out together? Can you remember the last time you sat and played a board game with the family? What was the last book you read? Is it really eighteen months since you visited your 'best' friend? There are all sorts of rituals that slow us down and allow us to take time out from our hectic schedules. They are especially rewarding, because they allow us to deepen our relationships with one another. Just think how much more enjoyable mealtimes would be if they were less frantic and you didn't have to compete with the TV to get your partner's attention. You could follow a meal with a games evening for the family, for example.

You can also create home rituals for yourself. Why not sit down to a favourite film or look through old photo albums? You could also consider taking up a hobby you enjoy. Or, you could block off some time for a 'home spa' and give yourself some relaxing beauty treatments.

Nostalgia

When you think back to your own childhood, was there anything that your parents or grandparents did around the home that you enjoyed as a child and that you could do for your own children? You could make a ritual of decorating the home for a special occasion or religious festival. Or why not get together with some neighbours and organise a street party?

Share the burden

Despite surrounding ourselves with a range of expensive labour-saving devices, we're spending more time than ever before on household chores. The average British family now washes five loads a week, which uses large amounts of energy, water and detergent; this harms the environment and saps our cash and time.

Do you remember that granny used to have a 'wash day'? And granny never seemed to be stressed and rushing around like we are. For example, even though our washing machines and tumble driers mean we can do a lot more laundry a lot more quickly, we just do more and more of it.

Unlike granny, more women work these days, yet women still tend to be the ones who do most of the household chores. Sharing these chores with the family will ease the burden on poor old mum. Why not draw up a list of chores that need to be done and mark on there who's responsible for what? You could even link it to the kids' pocket money. Again, it also acts as a useful time audit, so when you look at the list you can ask yourself whether it really is necessary to do eight loads of washing a week.

All change

Throughout this book, I've been trying to convince you that you need to change some of your habits. However, change can also be quite stressful and while the end result may be beneficial, the process could be pretty traumatic for some of you.

A lot of the stuff you need to do to cut down on your waste means casting off some emotional and habitual baggage that you may have carried with you since childhood and to start to look at things in a new way. You may look back and remember how your own parents dealt with household waste, for instance. Did that influence the way you think about what you waste and what you throw away? How many times can you remember your parents saying to you, 'Isn't it about time you got rid of that old thing?' or 'Chuck that out. It's no good. It's rubbish.' If you've been brought up to behave in a certain way and to believe certain things, then changing what you know and what you're used to doing can be extremely difficult.

Don't be too hard on yourself if you're finding some of this difficult to deal with – start small and make a few simple changes. Soon you will be able to appreciate the bigger picture and, ultimately, start thinking more about the planet as well as saving time and money. The more you experiment with making some small changes, the more confidence you will have to carry on and try new things.

Remember the saying, 'Old habits die hard'? Well, many psychologists believe that it takes between three and six months to change a habit and to change our behaviour. Therefore, relapses could be on the cards. For example, you may find yourself so busy at work one week that you grab those ready-meals in their wasteful packaging and hey presto – your bins are full to overflowing again. At this point, it's going to be hard to pick yourself back up again, but whatever you do – please, don't give up. It's at times like these that you realise how important it is to carry on and not to throw in the towel. Your actions, however small (or sporadic) they may be, really do make a difference.

TOP TIPS TO IMPROVE YOUR HAPPINESS AND SAVE YOU TIME

CHOOSE HAPPINESS

As Abraham Lincoln once said: most people, most of the time, can choose how stressed or happy, how troubled or relaxed they want to be. So, choose to be happy!

Watch less TV. Watching excessive amounts of TV is bad for our health – especially for our children. There's more childhood obesity than ever before, which increases the risk of our kids developing heart disease and diabetes in later life. Instead of letting the children slob on the sofa staring at the gogglebox, encourage them to swim, dance, play sports or read. As well as reducing their expectations with regard to whatever the latest craze happens to be, they will be much healthier.

TAKE TIME TO DO WHAT IS USEFUL AND REWARDING

Plan your time wisely and if you feel like being spontaneous – go for it! But make sure you set aside time in your planner to be selfish and to do what you want to do, when you want to do it. Use your time wisely. Eat when you're hungry, nap when you need it. Don't let the pace of life dictate how you use your time. If you want to enjoy a more leisurely feast, then plan a family meal, a dinner party, or have a 'clothes swap' party with some girlfriends – get them to bring along some food to share the cost and preparation.

MAKE YOUR RELATIONSHIP A PRIORITY

Show your love! Let your family and friends know that you value them, by making time for them.

LEARN SOMETHING NEW, EVERY DAY

To be happy, most of us must also be growing, expanding, learning and challenging ourselves. Read, listen, adapt and stretch to accommodate new ideas and new information.

USE YOUR BODY AS IT WAS DESIGNED TO BE USED

Walk and run, stretch, throw things, and lift things. Dance! Sing! Jump about and generally have a blast! Exercise is good, but making love with your partner is great! Making fairy cakes with the kids, cuddling your best mate, laughing with your friends and generally having a good time are brilliant stress busters. You've got a body, so use it! It can either be a source of pleasure and fun, or a source of aches and pains. Avoid toxins. People can also be poisonous. Give all the energy vampires and negative creeps the elbow and avoid poisons whenever possible.

KNOW WHAT MAKES YOU FEEL ALIVE AND CREATIVE

Try making cards for birthdays, and other special occasions. They will mean more to the recipients and unleash your creativity.

TRY NOT TO MAKE PROMISES YOU CAN'T KEEP

It's too easy to be harassed and you could find yourself giving in to every demand made on your time. Learn to say 'no' and mean it.

PAY KIDS FOR HELPING WITH EXTRA CHORES

Apart from freeing up your time, this allows your kids to gain some extra money, which they can put towards something that they want.

Write a list of things you would like to achieve in the next five years and work towards them . . .

What you have saved . . .

SAVED		SO TREAT YOURSELF . . .
A potential £200 from selling your junk: old records make a tidy sum on eBay, electrical items can be sold in *Loot* and everyone can make something from a car boot sale.	→	Give something back – invest in a charity project or sponsor a child.
Time and lots of it – more efficient time-planning could save you and an hour a day, seven a week, 365 a year, etc.	→	Write that inspirational novel that you've been promising yourself you would do for ages.
Your sanity – make way for peace and quiet in the home.	→	Stop making crazy choices and start realising that life is too short!

If you just do one thing . . .

Eat meals together.

Nowadays we tend to spend an average of 20 minutes cooking a meal and 10 minutes scoffing it down. What sort of conversation can you have with someone in 10 minutes?

CHAPTER SEVEN

Thinking outside the box

I want you to take a step outside your front door and take a good deep breath . . .

If you live by a busy road then no doubt that deep breath will have resulted in getting a big whiff of vehicle exhaust fumes.

What about your garden? Do you apply the three 'Rs' in the outside space you're responsible for?

DRIVING CRAZY

Air pollution from motor vehicles has become a major problem – especially for our health – and it is set to worsen. Most vehicles use petrol or diesel made from oil and their exhaust gases go into the air that we breathe. Smelly fumes, noise, traffic jams, endless rounds of diversions and road works, kids screaming in the back and increased blood pressure are all part of our everyday routines. Isn't it strange how much we're prepared to put up with, simply because we won't give up our cars?

Before you throw this book to the floor in disgust – I'm a realist, so I know that I'm not going to coax you all out of your cars to don a pair of eco-sandals and set off on a 20-mile route-march to work every day. But, I want to persuade you to think about how you could use your cars (or motorbikes) a bit differently and to think about alternative forms of transport.

Smoke signals . . .

Of course, air pollution from increased traffic use isn't just from us jumping into our cars at every opportunity we can get. If you join the dots and look at how you shop, what you buy and what you throw away, you'll see that all these things have a direct link to increased road usage. We're becoming more and more reliant on road transport for moving all the goods we buy – and if we're causing more pollution because of the goods we're demanding, then it's up to us to cut our consumption.

Just look at what the fumes produced by road traffic do to our health:

Carbon monoxide –
reduces the amount of oxygen carried in the blood, causing headaches and vomiting

Diesel smoke –
can cause cancer

Lead –
damages the central nervous system and can cause brain damage in children

Nitrogen oxides –
irritate the throat and eyes

Sulphur dioxide –
causes coughing and a feeling of tightness in the chest

Benzene –
causes anaemia and is linked to leukaemia

Scary, isn't it? And even more so when, according to the National Travel Survey, the most healthy form of travel, walking, has slumped by 20% in Britain since the mid-1980s, whereas car usage has shot up 40% in the same time period. Don't forget, just because you're inside your car it doesn't mean you're escaping the fumes, so crank open the window and get some fresh air.

AUTOGEDDEN

- People in the UK make more use of cars than any other European country, despite having below-average car ownership. (Source: CfIT 2001)

- Around 85% of households in rural areas have at least one car, compared with 70% of households in urban areas. Across Britain car availability is lowest in London, with 63% of households owning a car, and highest in the Southeast, where 80% of households have at least one car and 33% of households own two or more. (Source: National Travel Survey)

- Children living on heavily trafficked streets are more likely to develop chronic respiratory problems. (Source: *Occupational and Environmental Medicine*)

- In the UK, 84% of people hear traffic noise at home and 40% of people are bothered, annoyed or disturbed by it; 28% of people say that road traffic noise at their homes has got worse over the past five years. (Source: DEFRA)

- Between 1985 and 2002, 20,000 hectares of land have changed to transport use from previously undeveloped land, roughly equivalent to an area three times the size of the urban area of Nottingham. (Source: DfT Transport Trends 2002)

- Men cycled more than women in Britain in 1999/2001, making 23 journeys per person per year overall compared with nine journeys for women. (Source: National Travel Survey)

- Women used taxis more than men in Britain in 1999/2001, with 13% of women and 11% of men using a taxi at least once a week. (Source: National Travel Survey)

- Motorcycles are more polluting (in terms of emissions per passenger kilometre than cars) for particulates, carbon monoxide and volatile organic compounds. They are less polluting than cars for nitrogen oxides and sulphur dioxide. (Source: Strategic Rail Authority)

- A worrying 60% of men and 70% of women are so physically inactive that they risk coronary heart disease, diabetes, stroke or obesity. (Source: Health Education Authority)

- The health impacts of traffic pollution cost £11.1billion each year. (Source: Environmental Transport Association)

- A double-decker bus carries the same number of people as 20 fully laden cars. (Source: TravelWise)

ECO-ENGINES

- The more fuel that is burned, the more greenhouse gases are produced, so you could opt for a car with a smaller, less polluting engine.

- Many cars can be converted to run on liquefied petroleum gas (LPG), or can be made to run on dual fuel (a mixture of LPG and petrol). However, LPG isn't cheap when compared to petrol, so a more economic alternative could be to convert a car to run as a hybrid.

- A hybrid car has a petrol engine and an electric motor; the electric motor's batteries are charged when the car is braking, decelerating or cruising. That means you don't need to recharge it at a power point. There are already a few hybrid cars on the market in the UK and the biggest advantage is that their fuel consumption is as much as 30% lower than equivalent petrol cars.

- Wholly electric cars produce no emissions on the road, have lower fuel costs than their petrol equivalents, and maintenance is cheaper, too, because they have less moving parts to service and replace. They can increase emissions at the power station though, unless you're on a green electricity tariff (see Chapter 3). To refuel your electric car you simply need to plug it into the domestic mains supply for around seven hours. Some city centres now provide free parking and recharging ports for people using electric cars. The beauty is that they can be recharged from virtually any standard 13-amp socket anywhere, and to keep costs down you can take advantage of off-peak rates.

- There are clearly environmental benefits to using different types of fuel and vehicles, but as these are still fairly infant technologies there are a few practical disadvantages, which is why most of us choose to stick to petrol or diesel cars. Regardless of these disadvantages, you can still be more environmentally friendly by choosing a car with lower emissions and being aware of how you use your car.

LEANER, GREENER MOTORING

Before you make any journey you should ask yourself if it's really necessary to use the car. The shortest journeys (those less than two miles) are the biggest polluters because it often means our engines are cold and are straining to warm up – which means they use more fuel and create more pollution. It's all about using your common sense (and sometimes your legs) instead of using your car for every trip.

- Set a weekly target for trips to work or for social activities using the bus, train, walking or cycling, instead of using your car. Try to increase this each month to ease yourself in gently. If you stick to your targets you'll soon feel the benefits of your greener, healthier lifestyle.

- Get on your bike! It's good for your health and bikes are cheap to buy and maintain. What's more, they're free to use and park.

- If you don't fancy cycling, why not use public transport to get to the shops? Most of the larger supermarkets have bus stops in their grounds, so you don't have to struggle down the road to catch the bus, you can just hop straight on and off with your shopping.

- Plan your journeys by checking public transport websites (e.g. www.nationalrail.co.uk) and by ringing local public transport providers for timetable information and fares. Ring local licensed taxi firms to get quotes and display 'car share' notices at work or, if you have children, at the school.

- If you have kids, get them on their feet. Encourage them to walk or get them to cycle where it is safe to do so. A rota of parents could take it in turns to walk the kids to school. Or if cycling appeals, then visit Cycle Training UK's website www.cycletraining.co.uk – they are the UK's biggest independent provider of on-road cycle training, and they also provide cycle maintenance training in London. They train instructors throughout the UK, so are a good place to start to look for local instructors. They also supply tailor-made training for individuals and families.

- Plan combined trips, such as work/shopping and walk or cycle to your local shop for 'top-up' items, like bread and milk.If you really can't pull yourself away from your car, could you share a car journey with a friend or colleague?

- Apart from being dangerous, aggressive driving causes more pollution, which means higher fuel costs. Try to avoid rapid acceleration and heavy braking. Smoother and steadier driving can use 30% less fuel. And obey the speed limit – if you don't then you're not only wasting fuel but you face a hefty fine or even a prison sentence. Just think – driving at 50mph can use 25% less fuel than at 70mph.

- Sitting in your car with the engine idling is a big 'no-no', as you're pumping out loads of pollution and it wastes fuel too. If you have a catalytic converter – which is a device that removes pollution in the exhaust – then sitting in your car with the engine running stops the catalytic converter from working properly, so don't do it!

- Make sure you get your car serviced regularly at a garage – usually every 12 months or every 10,000 miles. A well-maintained car can use less fuel, so keep an eye on how much fuel you're using as this can be a good indication that your car isn't working as efficiently as it could. Under-inflated tyres, for example, can increase fuel consumption by up to 8%.

- If you're not using them, remove any roof racks from your car as they increase the wind resistance and that means more fuel use. Keeping unnecessary bulky items in the boot also adds to the weight of your car and increases fuel consumption.

- Do you really need that second car? Why not buy a motor scooter, moped or small motorcycle instead? These are cheaper to buy and run, reduce traffic congestion and produce fewer emissions than cars.

- Know what you're buying and make smarter car choices. If you're buying a new car, go for the smallest and most economical possible. The Environmental Transport Association (ETA) Car Buyer's Guide assesses cars according to their environmental impact and is available from the ETA for approximately £5. Visit their website (www.eta.co.uk) for loads of green motoring tips. Wearing the right footwear (flat and comfortable) helps you to control the accelerator and brake, as you'll be more sensitive to the pressure you need to apply. Abrupt braking and accelerating increases harmful emissions. So, think about what you've got on your feet and consider buying driving shoes if you haven't got anything suitable in your wardrobe.

Aggressive driving causes more pollution, which means **higher fuel costs.**

HOW DOES YOUR GARDEN GROW?

Whether or not you're lucky enough to have a garden, there are plenty of things you can do to cut down on waste and become a bit more 'green' fingered.

First up, if you really want to cut back on your supermarket shopping, then why not grow your own food? Growing your own food can take time, but shared allotments are on the increase and mean that when you go on holiday, someone else can water your plants.

You don't have to own half of Hampshire if you want to grow your own food. If space is limited, or if you live in a flat, you'll be surprised at just how much you can grow on a windowsill or in containers. Sprouting beans, peppers, chillies, tomatoes and strawberries can all be grown on a windowsill.

Herbs will need regular feeding and watering and you must be prepared to change them regularly. Hanging baskets can also be used to grow them, so that may be an alternative for you.

If you live in a flat, container vegetable growing is a really good way to grow your own food. If space is limited, go for climbing plants, such as runner beans or peas, which use up less space. Containers dry out quickly on sunny days, so you'll need to keep them well watered and fed. Don't waste time and effort on root crops like beetroot, turnip or things such as cabbages – instead go for vegetables with shallow roots such as lettuce, tomatoes or spring onions.

For more information, the Henry Doubleday Research Association provides a whole host of information on growing organic food on their website www.hdra.org.uk, so check it out.

Good wholesome, home-grown food is one way of ensuring that we're not filling up our bodies with pre-processed muck

COMPOST: PEELING IT OFF AND GETTING MUCKY

GETTING STARTED

To help you take your first steps towards complete composting confidence, you could consider becoming part of a community composting scheme. Or you could contact the Waste & Resources Action Programme (WRAP) who run a home composting hotline – 0845 600 0323. This can be a useful first port of call for all your home composting queries.

WHERE TO BUY YOUR BIN

Because many local authorities now provide subsidised home compost bins, it's probably best to start by calling them first to see if they can provide you with one. Alternatively, you could try any of the following suppliers:

- The Organic Gardening Catalogue – www.organiccatalog.com – supplies a good range of compost bins.

- The Bin Company not only sells home composting bins, but also biodegradable bags, and wormeries. See their website www.thebincompany.com or call them on 0845 6023 630.

- The Green Cone converts organic kitchen waste into 90% water, oxygen and a small amount of soil conditioner, which seeps in to the surrounding soil. Made from 50% recycled plastic it costs £59.99 plus £6.99 postage and packing. This might seem like an expensive option, but some local authorities offer the Green Cone, so it's worth checking. Call 0115 911 4372 or visit www.greencone.com for more information.

- Recycle Works have a wide range of domestic compost bins in various styles (many are made from FSC timber), plus wormeries and shredders. Call 01254 820088 or visit www.recycleworks.co.uk for details.

WHERE TO PUT YOUR BIN

Compost bins don't normally have, or need, a base as free access to the soil allows good drainage. Being plonked straight on to the earth also means that worms and other insects that help the composting process can get in. However, you'll still need to put your compost bin somewhere where it's easy to use.

HOW LONG DOES IT TAKE TO MAKE COMPOST?

Compost can be made in six to eight weeks, or it can take a year or more. You'll know when it's ready because all the stuff you've put in the compost bin has turned dark brown. It should also smell slightly earthy, but it's still best if you leave it for a month or two to 'mature' before using it. Don't fret if your compost is a bit lumpy or stringy, with bits of eggshell and twigs dotted around; it's still perfectly usable.

When you finally get around to using your compost, simply lift the bin up in the air (you may need two of you to do this) and turn it out. If you've got the 'balanced diet' bit right then your compost should stay like a big neat sandcastle. You then need to take off the top un-composted layer (which was sitting at the bottom of the bin) and the wormy layer just below that and use this material to start your next binful. The remainder is what you can dig into the top 150 mm of your soil, or use in your planters.

WRIGGLY WORMS

A lot of compost-converts find that using a wormery helps with their composting. A wormery is divided into a number of chambers and one of these houses a number of worms. It's dead simple to use, as all you have to do is drop your daily kitchen waste into the bin and forget about it. The worms feed on the food waste and convert it into concentrated liquid feed and bio-rich organic compost, which you can then dilute with water and use as a plant food.
Visit www.wigglywrigglers.co.uk and www.originalorganics.co.uk

The Henry Doubleday Research Association (HDRA) is THE place to find out in greater detail how to make the perfect compost. The National Organic Gardens in Ryton, near Coventry, is home to HDRA and a great place to visit and be inspired to compost and grow your own. Their website contains lots of fact sheets, check them out at www.hdra.org.uk

Making compost is a bit like making a cake

you need the right ingredients, the correct temperature and moisture

WHAT TO TOSS IN THE COMPOST HEAP

You'll need to feed your compost a good mix of 'greens' and 'browns'. Greens are good 'activators', which mean that they rot quickly so will help to get the composting process going. All your fruit and veggie peelings and uncooked fruit and veggie leftovers, weeds, young hedge clippings, crushed eggshells, tea bags and coffee grounds, dead flowers and just a dash of grass clippings are ideal activators! Browns are all the things that cook a bit more slowly. These materials are needed to give body to the finished compost – otherwise you'll end up with a big, smelly, slimy mess. To get things to rot more quickly it helps if you chop everything up beforehand. This also reduces the bulk of your compost dramatically, so you can get more in when you need to. There are also super ingredients that heat up and activate your compost.

ADD THE FOLLOWING

Fruit and vegetable scraps
Tea bags
Coffee grounds
Old flowers
Bedding plants
Old straw & hay
Vegetable plant remains
Strawy manures
Young hedge clippings
Soft prunings
Perennial weeds
Gerbil, hamster & rabbit bedding
Wood ash
Cardboard
Paper towels & bags
Cardboard tubes
Egg boxes

SUPER INGREDIENTS

Comfrey leaves
Young weeds
Grass cuttings
Chicken manure
Pigeon manure

SLOW TO ROT

Autumn leaves
Tough hedge clippings
Woody prunings
Sawdust
Wood shavings
Crumpled cardboard
Plant stems
Twigs
Egg boxes
Animal manures

DON'T PUT IN

Coal & coke ash
Cat litter
Dog faeces
Disposable nappies
Glossy magazines
Plastic
Glass
Metal
Perennial weeds such as dandelions
Diseased plants
Meat
Fish
Cooked food

(source: Henry Doubleday Research Association Website
www.hdra.org.uk/organicgardening/compost)

WATER WASTING IN THE GARDEN

Saving water applies as much to our gardens as it does to our homes. Our gardens need water most when it's summer, which is why you'll often see a whole neighbourhood out with their watering cans, sprinklers and hose pipes merrily dousing their gardens with water. Most people tend to use tap water – this has come from our rivers, streams or aquifers, which are underground rocks that supply rivers, streams and wetlands with water throughout the year.

If you look at the knock-on effect our waste can have not just on the environment, but on our purses, a shortage of water means we're having to pay more and more for it. Water companies pass on the costs of treating water to their customers either through metering your water supply or through standing charges. So, excessive watering in the garden is costing you! There are quite a few simple tricks to save water in your garden:

Recycle water from your home and use it in the garden

Improve soil structure and use materials to keep moisture where it's needed most and for longest

Use various irrigation systems instead of hoses or sprinklers

Grow drought-tolerant plants

Use containers and pots for smaller plants

GET YOUR PRIORITIES RIGHT

All plants need watering, but some are more resistant to drought and can cope without watering during short dry spells. Choose plants that are naturally more tolerant to dry soil such as lavenders, santolina and stachys. Drought-resistant plants, such as cistus, are also especially good for sandy soils.

KEEPING THINGS MOIST

Adding plenty of organic matter to soil before planting will help the roots of young plants to hold their moisture – another good reason to get composting. You can further keep moisture where it's needed by applying a mulch of compost, bark chips or cocoa shells to the surface of the soil. Mulches help to keep in water and keep weeds in check.

Spring is the best time to apply mulches but you can add them at any time as long as the soil's damp. Large containers can also be topped off with a mulch of bark or gravel.

HANGING BASKETS

Hanging baskets dry out very quickly, which is why, when you get around to watering them, water tends to pour through the bottom. That's because really dry compost repels water. Try to buy hanging baskets with built-in reservoirs to give roots a chance to benefit from watering rather than having it all run away through the bottom.

WATER IN THE EVENING

The best time to water your garden or plants is in the evening, so that moisture doesn't immediately evaporate in the heat of the sun. Watering at night will also give them a chance for a long overnight drink. An automatic watering system, which uses a timer on a tap, can be set to come on at any time, even when you're away. Water seeps out of small holes from pipes, delivering water to where it is most needed, so they're a good investment.

MORE INFORMATION ON HOW TO SAVE WATER

Gardening Which? can only be purchased through subscription, but has regular features on how to save water in the garden. They also publish useful fact sheets and reports. Visit www.gwfreetrial.co.uk for a month's free trial.

Environment Agency: produces some useful leaflets to help you to save water in the home and garden. Call 01276 454 445 or visit www.environment-agency.gov.uk for more information.

GREY WATER

One way of recycling water for use in the garden is to use 'grey water' (water that has already been used for washing or bathing). Using grey water can save up to 18,000 litres of water a year for each person, which represents 33% of household water use.

Bath water can be siphoned off into buckets using a hose – but this is pretty time-consuming and backbreaking. An easier alternative is to buy a diverter, which allows bath and shower water to be diverted to hose pipes or storage butts. In addition, your washing machine can be fitted with a diverter that directly empties all water into a pipe or bucket. There are a few golden rules with grey water recycling:

- Never reuse water that contains strong detergents, chemicals or household cleaning agents.
- Never drink grey water and do not use it on edible produce, i.e. fruit and veg.
- Never reuse bathing water when a family member is ill or using a topical skin treatment.
- Do not store grey water for long periods of time, as it will go rancid.
- Always store grey water separately from other water supplies.
- Do not repeatedly water the same spot with grey water; spread it around the garden.
- Ensure you protect mains water against contamination by backflow (in order to comply with the Water Supply Regulations 1999).

WATER BUTTS

Collecting water from roofs or any flat surface is a great way to save water for use outdoors. You can either get a single water butt or you can link several together to collect water during rainy periods. A water butt can provide enough water for the entire garden during dry spells and they come in a range of sizes. Make sure you get one with a well-fitting cover to stop debris such as leaves from falling in. Some also come with a child safety lid. Visit www.greenhousesupply.co.uk for more details.

If we all used water butts, we could help to avoid local flooding. When we experience a heavy rainfall, the run-off from roofs, roads, pavements and other non-porous surfaces means that a torrent of water is sent to drains. When the system can't cope, you get flooding. When planning hard surfacing in your garden, or redoing your drive, consider using a porous surface to do your bit to reduce flooding.

TOP TIPS APPLYING THE 3 Rs TO THE GARDEN . . .

- Reuse some old packaging like bubble wrap – it helps to protect plants and containers against frost.
- Don't throw out those old net curtains just yet – they're great for shading areas of the garden and are good for keeping carrot fly at bay.
- Old car tyres make excellent planters for potatoes – stack them on top of each other as the potatoes grow.
- Fill an old pallet with compost to make a herb garden.
- Ice-lolly sticks can be used for labelling up your plants.
- If you're like me and forever dropping plates and cups, all your broken bits and pieces of crockery can be used in the bottom of your planters to help with drainage.
- Old kitchen foil, milk-bottle tops or all those embarrassing old 80s CDs can be strung together to scare the birds away.
- If slugs are your garden nemesis, reuse small plastic containers as slug traps. Fill the container with beer or milk and sink into the ground. Or if you'd rather be more humane, try scattering broken eggshells around your plants – they don't cross those. An old piece of carpet or a sprinkling of gravel should have a similar effect. They don't like coffee grounds either.
- Large plastic bottles make great mini greenhouses and keep slugs at bay
- Yoghurt pots are great for seedlings – just pierce the bottoms for drainage
- Encourage birds into your garden to help keep pests in their place by making a special area for wildlife and by choosing plants that provide food for birds and insects.
- Build a butterfly table by hanging a piece of plywood to a tree branch with a piece of string. Place jam-pot lids of sugar dissolved in water on top for the butterflies to sip, or plant buddleia and honeysuckle to encourage butterflies into the garden.
- Cut the necks off plastic bottles to concentrate watering where you need it – this is especially good for small border plants and anything in long planters.
- Use old egg boxes as seed trays.
- Keep a pile of woody prunings and hedge clippings in a corner of the garden to encourage toads and hedgehogs, who act as natural pest controllers.
- To fill in gaps in your lawn, sow grass seed on used tea bags and use them to fill the holes.

Little Marvel
pea.
5/2/05
L5

What you have saved . . .

WHAT YOU HAVE SAVED . . .	SO TREAT YOURSELF . . .
£2900 – the average household spend on motoring a year.	Buy bicycles for you and your family, or for a family in Africa. (www.goodgifts.org)
Your lungs and heart.	Simple: be more fit, more active and live longer.
£100 a month spent on fresh vegetables that you could have grown yourself in an allotment.	Buy a wormery for a wriggly way to speed up your home growing.

If you just do one thing . . .

Give a friend a lift. Whether it's a school run, taking an elderly person to the supermarket, or picking up your mate when they're stuck in the rain, get filling the wasted seats in your car. If you have to drive, make it worthwhile.

CHAPTER EIGHT

Taking things to the next level

With the price of energy and water continuing to rise, it has never made more sense for the average householder to invest in their homes, making them more energy and water efficient.

It makes sense to live as greenly and as leanly as possible and here's how you can save even further . . .

When my husband Gil and I bought our current home, we set about making it the greenest, cleanest and healthiest house in the street. And while we knew it was never going to be easy (or cheap), we were on a mission!

At the time all our friends and family thought we'd gone stark raving mad. There was no doubt about it, we'd bought a five-bedroom Victorian thermal slum complete with a damp, dodgy interior, mangy carpets, hideous 1970s kitchen fittings and little more than mould and general decay holding the bricks in place, instead of the usual mortar. We stripped the house of its hideous 1970s decor and used natural paints and finishes to create a healthy, beautiful interior. We scoured skips and second-hand shops for rugs and furniture. We were also given furniture by friends and family. When we did have to purchase larger items such as beds and sofas, we ensured the wood was from managed sources and that fabrics were unbleached.

Gil and I both agreed that our first job would be to make our home as thermally efficient as possible, and that we'd use our home as a laboratory to test new sustainable materials and products. Our ultimate goal, though, was to make the house as energy efficient as possible. We insulated the roof, dry-lined the front elevation and insulated the cellar boards with sheep's wool, a breather membrane and eco-board. And then we externally insulated the side and rear elevations. We'd soon installed a rainwater collection system, low-flush toilets and a solar hot-water heating system.

Of course, all of this costs money, so why on earth would anybody choose to go down this route? Well, for us, it's a question of principle. We wanted to practise what we preached and show people what's possible. It was a chance to take on the biggest recycling project of our lives – to transform a leaky old house into the most advanced eco-retrofit in the UK.

We don't expect everybody to follow our way of living, but visitors have said that it's motivated them to consider 'greener' options when they're planning various projects around the home and try things out for themselves.

If you really fancy taking recycling to the extreme – then why not recycle an old house?! We've invested a great deal of our time, energy and money in our home, but the old girl looks after us in return. So, if this has inspired you rather than scared you, you're probably ready to move on to the next level. You'll be surprised what you can do in your own home for a small sum, and the good news is that you may be eligible for a grant to offset some of the costs. Check out our website for further info www.msarch.co.uk/ecohome.

You'll be surprised what you can do in your own home for a small sum, and the good news is that you may be eligible for a grant to offset some of the costs.

RENEWABLE ENERGY

Using renewable energy sources such as solar and wind power can reduce the amount of fossil fuels we use which, in turn, cuts down on harmful greenhouse gas emissions that cause climate change.

The most common methods of harnessing renewable energy to generate electricity are wind, hydroelectric turbines or photovoltaic solar panels. Solar panels will give you decades of service, reduce your CO_2 emissions and cut your bills, but generating your own electricity is usually expensive. Using renewable energy to heat your home and to provide you with hot water, though, is much more cost effective.

HOT WATER AND A SUNNY DISPOSITION

Solar hot-water heating uses heat from the sun to provide up to 55% of your domestic hot water. There are two main types of solar water-heating panel available – flat plate and evacuated tubes. The cost of a flat-plate system, including the cost of professional installation, is from £2,000–£5,000 with a payback period of between 5–7 years. Evacuated tube systems are more expensive – ranging from £3,500–£6,000 for professional installation. However, prices do vary so it's worth shopping around to get a few quotes before you hand over your cash.

If that's too daunting, you can opt, like Gil and I did, to have it installed at a reasonable rate by a recommended contractor.

Once you're up and running, the system should need very little maintenance and should last for decades. An annual inspection of the panels and the system to see that everything's still working properly and that there's no corrosion or condensation should be all you'll need to do.

RAINWATER EQUIPMENT

There are lots of companies that provide rainwater equipment such as the one in our house which collects rainwater from the roof and stores it in large tanks in the cellar to be used to flush three ultra-low-flush loos, supply the washing machine and the outside tap. Contact CAT's free information service for details (email: info@cat.org.uk). They also publish a 'Water Supply, Treatment and Sanitation' resource guide, which is a comprehensive directory of useful contacts, and tip sheets including 'Water Conservation in The Home' and 'Making Use of Grey Water in The Garden'.

PHOTOVOLTAICS

Photovoltaics (PVs) also use sunlight to generate electricity, but they are expensive. They can cost an average of £10,000, but you may be able to get a grant towards some of the costs of installation.

For further information on choosing, setting up and maintaining a small-scale photovoltaic system, the Centre for Alternative Technology (CAT) publishes a resource guide listing manufacturers, suppliers and installers of PVs. (www.cat.org.uk)

Grants

There are a number of grants available to help make your home as energy efficient as possible. You could get between 40 and 50% towards the cost of installing solar electricity equipment. For a domestic user you could get up to £400. Check the grant search scheme on the Energy Saving Trust website (www.saveenergy.co.uk/gid) for eligibility.

WOOD-BURNING STOVES

Domestic wood-fired heating systems come in all shapes and sizes and can use various types of fuel, from logs to pellets. A wood-burning stove usually just heats one main room, and normally has a back boiler to heat water.

A domestic wood-burning stove will cost you around £500–£1,000 depending on the size and model you choose. Automated boilers are more expensive, but, again, you may be able to get a grant towards the costs of installation.

ECO-LOOS

This really isn't as basic as it sounds and composting toilets don't use any water – so that's a big plus. You can use the composted waste in non-food areas of the garden. Keeping urine separate from stools is key to a good composting toilet system, otherwise things can get a bit whiffy. There are various models available and can cost anything from a few hundred to several thousand pounds. You can also buy a dual-chamber compost toilet designed to fill up over the course of a year. The seat can be switched between chambers, so that the second is filled while the first composts down.

If you can't bear the idea of composting toilets you could fit an Aquatron – for about £500 – to the outlet of a flush toilet. This device separates solids and liquids, with stools going straight into a composting chamber and your pee being treated in either a leach field or reed bed. The toilets in my house are by Ifo (a Swedish company) and are ultra low flush (2–4 litres). Contact The Green Shop or Elemental Water Solutions for further details. I also have a composting chamber in my cellar. This is essentially a giant worm composter which gradually breaks down the waste in compost.

HAY, GOOD LOOKIN', WHAT YOU GOT COOKIN'?

Cooking with a hay box (or a Victory Oven) is an extremely energy-efficient method of cooking and literally involves lining a box with some hay and using it to cook your food. It's particularly good for cooking stews, casseroles, soups, rice, root vegetables and porridge.

You need a heavy dish or pan with a tight-fitting lid and you'll also need to heat food first on the hob for about ten minutes to bring it to boiling point. Once it's hot, you wrap the sealed dish in a towel and sit it in the hay box – rather like a nest. It takes about four times as long to cook food in a hay box than it would in a conventional oven. How quickly the food cools depends on the temperature outside the box, so it needs to sit in a warm place. And don't leave it anywhere that mice might see it as a potential des-res!

Food tends to cook better if you don't overfill the dish and it's important not to get the hay wet from either the steam or from spilling any liquid.

If you look after it, the hay can be used again and again, although you may need to add a little more as time passes, as it will tend to bed down. If you're not planning to use your hay box for a while, then make sure the hay in it is dry. Wet hay develops microbial activity which then increases the temperature of the hay and can cause fire. That's why hay is normally stored above animals in barns!

The other added bonus of a hay box is that you can make your own out of recycled materials you have lying around, so it's a great way to reuse and recycle household waste. The Centre for Alternative Technology provides an excellent fact sheet on how to make and use a hay box. (www.cat.org.uk)

'GREEN' PAINT

Manufacturing paint uses loads of energy, producing up to ten tonnes of waste, much of it toxic. Add to that the toxicity of the paint itself and you can see why I don't think it's very pleasant stuff.

We've already seen that volatile organic compounds (VOCs) can irritate the eyes, nose and throat, which is why most paints now carry a VOC rating. Try to go for the lowest possible VOC rating you can. However, some of the water-based paints low in VOCs contain even more harmful chemicals:

- **Alkyl phenols** – these are hormone disrupters which also bioaccumulate.
- **Vinyl resins** – can cause lung and liver damage as they're known carcinogenic skin irritants.
- **Titanium dioxide** – often used for improved whiteness or opacity, it is harmless to use, but the amount of energy used in its manufacture is an environmental concern.

A number of paint manufacturers now specialise in 'eco' or 'natural' paints, using natural, biodegradable ingredients. Some paints even use orange peel, which makes them smell good enough to eat (but please DON'T as they still contain poisonous materials). Natural paints can cost more than conventional paints, but the hidden costs to your health and to the wider world will be much less.

Visit the Green Shop (www.greenshop.co.uk) or Natural Building Technologies (www.naturalbuildingproducts.co.uk) for a range of products.

It may still be worth asking for a list of ingredients, though, as some 'natural' paints still contain synthetic alkyds, white spirit, vinyl resins and titanium oxide, which you may not want to use.

If you really can't afford a natural paint, then try to use up cast-offs that would otherwise go to waste. At the very least, try to make sure that any spare paint you have lying around goes to someone who needs it and will use it.

TOP TIPS GREENING YOUR HOUSE

This is mine and Gil's list of recommendations for the most common eco-home improvement projects:

- The first step is always to reduce the amount of energy that you use. Install as much insulation as you can in the roof and make sure drafts are minimised. Our website can tell you how we did it (www.msarch.co.uk/ecohome).
- Consider installing a solar hot-water heating system. It's getting cheaper all the time and only basic plumbing skills are needed for most installations.
- Sourcing eco-alternatives to nasty stuff like fibreglass insulation are becoming more common. We used Warmsell in our home, which is a highly effective material made from recycled newspapers (www.excelfibre.com), sheep's wool from Second Nature (www.secondnatureuk.com) and external wall insulation from Sto. (www.sto.co.uk)
- Check out non-PVC alternatives to piping and cables – it could be phased out in the UK in a few years, following on the heels of an increasing number of European countries.
- Instead of daubing your walls in petrochemical-based paints, try more natural alternatives that are kinder to the environment and look fantastic too.
- If building cupboards or shelving look for sources of reclaimed timber. It's often of superior quality and very competitively priced – and it has none of the brassiness of younger wood.
- Next time you buy a light bulb, make it a low-energy one – and scoop savings of up to £50 over the life of an eco-bulb.
- Switch to a green electricity supplier – you may not even have to change your current supplier. Give them a call to see what's on offer.
- Use natural waxes on floors; you'll find the wholesome smell quite addictive.
- When purchasing products, think about what they're made of and where they come from – the more natural the better, and the more locally they are sourced the less transport needed.
- Compost your kitchen and garden waste and improve your soil for free.
- Remember if you have a cellar to insulate under floors – up to 25% of heat can be lost this way.
- Check skirtings, windows and doors are draught-proofed.
- Install efficient heat-recovery fan units in wet areas to cut down on humidity and rapid air-cooling.
- Buy second-hand furniture and repair it – for cheery kitsch that you can alter by the season!

Start with small steps

and build good habits.

PUTTING IT ALL TOGETHER

The truth is that the only thing that stops us wasting less is that we're creatures of habit and, despite knowing the benefits of changing our behaviour, breaking a habit is never easy. Some of the changes outlined in this book do require a bit of effort and they almost certainly require you to think in a different way. The ideas I've shared with you are, of course, only suggestions, but the possibilities are limitless and it's up to you how far you take them. Not everything can or should be done right away and your own common sense will tell you whether or not you will be able to tackle some things now, or perhaps leave them for later.

At the start of this book I said that my life's ambition has been to get people to really use their brains and I hope that I've managed to convince you to use yours in a different way, so that you can reap the benefits of a healthier, happier and wealthier life. The rest, as they say, is up to you.

Here are a few final tips to get you started on the path to a less wasteful, time and cash rich, and all round more blissful existence.

Remember my lifestyle rules –

Reduce, Reuse, Recycle.

Watch your 'waste line' –

it's alarming just how quickly it spreads when we're not looking.

Convenience costs –

it may save you a bit of 'time', but it's a false economy.

Take your time –

accept that things can't always be perfect; there simply aren't enough hours in the day.

Go for 'green' –

before you buy, stop to ask yourself if you could do without it, that's really the 'greenest' choice you can make. If you really can't do without, then try to buy 'green' and ethical goods.

If you just do one thing . . .

The best thing about learning to save the planet is that the whole family can get involved. At the end of each chapter I've asked you to do just one thing to help cut down on the waste generated from your home, but now I'd like you to step up your work. Every time you or any of your family members do something from the chapters, mark it with one of these symbols. This is a great way to follow your progress and reward yourself when you've reached – and stuck to – your goals.

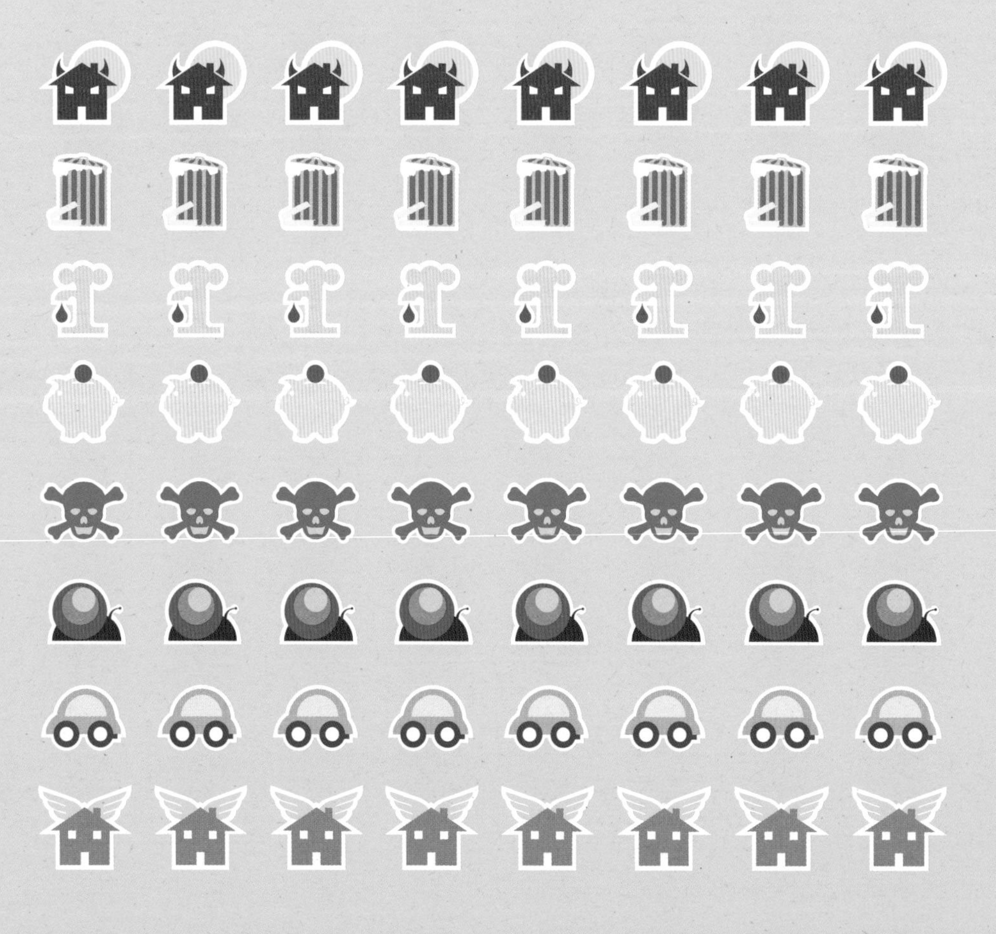

START ▶

FINISH

CONTACTS

The contact details below are featured in either this book or the *No Waste Like Home* television series. Websites and telephone numbers are correct at time of print only. Some companies are based in the television show's participants' home towns and will not necessarily provide services to the whole of the UK.

GENERAL

Advertising Standards Authority: www.asa.org.uk
Carbon Trust: www.thecarbontrust.co.uk (08000852005)
Cats Protection League: www.cats.org.uk
Canine Defence League: www.vetstream.co.uk
Centre for Alternative Technology: www.cat.org.uk
Classic Balloons: www.ballooning.fsnet.co.uk
Department for Environment, Food and Rural Affairs (Defra): www.defra.gov.uk
Energy Savings Trust (EST): www.est.org.uk (02072872087)
Ethical Consumer: www.ethicalconsumer.org
Fairtrade: www.fairtrade.org.uk
Forest Stewardship Council (FSC): www.fsc-uk.info
Friends of The Earth: www.foe.co.uk
Global Action Plan: www.globalactionplan.org.uk
Government Environment Agency: (08708506506)
Green Choices: www.greenchoices.org
Green issues: www.greenchoices.org
Greenpeace: (02078658226)
Green Shop: www.greenshop.co.uk
London Remade: www.londonremade.com
Natural Collection Catalogue: www.naturalcollection.com (08703313333)
National Statistics: www.statistics.gov.uk
Protective Red Suits: www.dickies-uk.co.uk
Robert Dyas: www.robertdyas.co.uk
SWAP (Save Waste And Prosper): www.swap-web.co.uk
Waste Online: www.wasteonline.org.uk
Waste Watch: www.wastewatch.org.uk
Watch UK: www.recycledproducts.org.uk (08702430136)
Womens' Environmental Network: www.wen.org.uk
WRAP: www.wrap.org.uk

BUILDING/PLUMBING

B&L Multi Trade: (02082450463)
Chelsfield Electrical: www.chelsfield-electrical.co.uk
Eco Housing Project (Bedzed): www.bioregional.com
Green Dream Consultancy: (01536 330564)
Gusto Construction Ltd: www.gustohomes.com
Hockerton Housing Project: www.hockerton.demon.co.uk
Ivor Skips: (02085756861)
KS Building Services: (01689876995)

RECYCLING

Bank Schemes: www.recycle-more.com
Batteries: www.rebat.com
Bins: www.homerecycling.co.uk; info@homerecycling.co.uk; www.binimizer.com; www.wrap.org.uk
Can Crusher: www.naturalcollection.com (0870 331 3333); www.homerecycling.co.uk
Civic Amenity Sites: www.wastewatch.org.uk (0870 243,0136); www.wastepoint.co.uk; www.wasteonline.org.uk
Electronic Equipment: www.nru.org.uk; www.computersforcharity.org.uk
Fridges and freezers: info@create.org.uk (01942 322271)
Furniture: www.frn.org.uk
Glass: www.milagros.co.uk
Junk Mail: www.mpsonline.org.uk
Metals: www.re-foundobjects.com
Oils: www.lessmers.co.uk; www.oilbank.org.uk (0800 663366)
Paint: www.swap-web.co.uk (0113 243 8777)
Plastics: www.recoup.org
Recycling aids: www.lakelandlimited.com

RECYCLED/GREEN PRODUCTS

Baby products: www.greenbabyco.com
Cleaning products: Ecover Cleaning Products, www.ecover.com (01635 574553); Green Shop, www.greenshop.co.uk; Earth Friendly Products, www.greenbrands.co.uk; www.dripak.co.uk
Clothes: www.greenfibres.com; www.hug.co.uk; www.motherhemp.com; Retro Man, 34 Pembridge Road, London, W11 3HL (020 7792 1715)
Cosmetics and toiletries: Weleda, www.weleda.co.uk
Furniture: www.greenworks.co.uk
General: www.co-op.co.uk
Nappies: National Association of Nappy Services, www.changenappy.co.uk; Cotton Tails Nappy Service, nappies@cottontails.co.uk (01244 374521); Kushies Nappies supplied by Perfectly Happy People Ltd (0870 1202-018); www.thebabycatalogue.com; Number One For Nappies, www.numberonefornappies.co.uk (01992 713 665); Cumfy Bumfy, www.cumfybumfy.co.uk; Women's Environmental Network, www.wen.org.uk
Paints: Association for Environment Conscious Building (01559 370908); Green Shop, www.greenshop.co.uk; Natural Building Technologies, www.naturalbuildingproductscouk.ntitemp.com
Product safety: Friends of the Earth, www.foe.co.uk; Greenpeace's Chemical Kitchen, www.greenpeace.org.uk

ENERGY SAVING

Air Conditioning: Ventaxia, www.vent-axia.com (01293 526062)
Boilers: www.boilers.org.uk
Draft Excluders: B&Q, www.diy.com
Energy Advice: Energy Saving Trust, www.est.org.uk; www.saveenergy.co.uk
Energy Advice Centres (EEAC): (0800 512012)
Fat Trap Containers: Anglian Water, www.anglianwater.co.uk (01522 341 000)
Flooring: Compi supplied by B&Q, www.diy.com
Gas: British Gas, www.house.co.uk; Corgi, www.corgi-gas-safety.com
Grants: www.saveenergy.co.uk
Green energy: www.greenpower.co.uk; Clear Skies, www.clear-skies.org
Hay boxes: Centre for Alternative Technology (CAT), www.cat.org.uk
In Sink Erator: Charles Pearce, www.charlespearce.com (0208 995 3333); www.insinkerator.com
Insulation: www.msarch.co.uk/ecohome; www.excelfibre.com; www.secondnatureuk.com; www.electricshopping.com; SolarTex[a] Thermal Blinds, www.thomassanderson.co.uk; Curtain Material www.scope.org.uk; Construction Resources, www.constructionresources.com
Jacuzzi Covers: Selective Covers Ltd (01845 522856)
Light bulbs: B&Q, www.diy.com; EDF Energy (0207 752 2266)
Pump Taps: Plumb Center, www.plumbcenter.co.uk (0161 9286383)
Radiator Reflector Panels: B&Q, www.diy.com
Rainwater equipment: Centre for Alternative Technology (CAT), www.cat.org.uk
Solar power: National Solar Club, www.est.org.uk/solar (0116 222 0222); All Systems Go (01776702340); Solartwin, www.solartwin.com
Thermostats on the Radiators: M&E Contracts Services (01908 502 461)
Water: www.water.org.uk; www.hippo-the-watersaver.co.uk
Washing Machine: www.dyson.co.uk
Windsave: www.windsave.com
Wood Burning Stove: Chase Heating Ltd, www.chaseheating.co.uk

FOOD

Chickens: www.omlet.co.uk
Farmers' markets: www.farmersmarkets.net; www.farmgarden.org.uk; www.boroughmarket.org.uk
Fish: Marine Stewardship Council, www.msc.org
Fruit and vegetables: Organic Delivery Company, www.organicdelivery.co.uk; Soil Association, www.soilassociation.org; www.abel-cole.co.uk (08452 62 62 62)
Growing your own food: Henry Doubleday Research Association,www.hdra.org.uk
Home environment: Asthma UK adviceline (08457 010 203); www.flowers.org.uk; Mariposa Alternative Bodycare (01273 242 925); www.geejaychemicals.co.uk; Weleda Toothpaste, www.weleda.co.uk

GARDENING

General: www.athas.org.uk; www.thebincompany.com (0845 6023 630); www.hdra.org.uk; www.greencone.com (0115 911 4372); www.recyclenow.com; www.recycleworks.co.uk (01254 820088); www.wigglywrigglers.co.uk; www.organiccatalogue.com; www.originalorganics.co.uk; www.wrap.org.uk; www.soilassociation.org
Allotment: One Tree Hill Allotments, One Tree Hill, Honor Oak, SE22 (020 8699 6099)
Composting: http://www.compost.org.uk/dsp_home.cfm
Composting courses: Ryton Organic Gardens, (024) 7630 8211
Future Forests: www.futureforests.com

TRANSPORT

Cars: Environmental Transport Association (ETA), www.eta.co.uk; Ecozone www.ecozone.com; Toyota Prius, www.toyota.co.uk (0845 275 5555); Dieselveg, www.dieselveg.com; Simple Green UK Car Wash, www.simplegreen.co.uk (01726 891199); The Environmental Transport Association, www.eta.co.uk; www.travelwise.org.uk
Cycling: Raleigh, www.raleighbikes.com; www.bikesandtrailers.com; www.cycletraining.co.uk
Rail: www.nationalrail.co.uk; www.railpassengers.org.uk

VIRGIN UNITE – INKJET CARTRIDGE AND MOBILE PHONE AMNESTY

Virgin Unite – the clue's in the name.
Getting involved and doing things differently, it's always been the Virgin way. So mobilising all we've got into a brand new charitable body – well, it just makes sense.

As well as raising money for some fantastic charities, please help us protect our planet by recycling your inkjet cartridges and mobile phones instead of chucking them away. It will cost you nothing and you will raise £3.50 for each mobile phone and £1 for each inkjet cartridge recycled.

Order recycling envelopes via our website www.virgin.com/unite or by calling the following numbers:
For mobile phone recycling call 0208 274 4040 quoting Virgin Unite
For inkjet cartridge recycling call 0800 435576 quoting Virgin Unite
(note that we cannot recycle Epson cartridges)

INDEX

ACKNOWLEDGEMENTS

It is a great pleasure to be able to acknowledge the love, help, expertise and support of the many people who have made this book possible. My thanks go to:

'Women's Hour' goddess Jenny Murray, who played an interview of me ranting on about my eco-home in Nottingham – Anna Richardson at Celador heard it and called me to see if I would be interested in becoming a presenter.

Love and thanks to my inspirational husband Gil – our work together made all this possible, to my mum who is my anchor and dear friend – I am as proud of you as you are of me. Lucy my gorgeous daughter, her partner Jay and my precious baby grandson Reeve – in some small way I hope this book contributes to a positive future my loves. Love to all my Schalom family – especially great-great Nono, Aba Nono, Guy, Dav and Louise and in loving memory of Nina Cherie. Big kisses to mum-in-law, Louise, Chilli and Doctor Art.

To have such friends as we do is to be blessed; special thanks to Lindsay, Spike my soul brother and Windy, Ferd and Josune, Nige Breeyo, Carl, Jen and Alfred, Ed and Bec, Fitz and Naomi, Gareth and Xanthe, Dom and Jay – the hosts, Dan Panorama, Goth Nick and Robin. And you others!

Special thanks to Ruth Turner – without her skills this book would not have come to fruition. Love and heartfelt thanks to my friend and agent Nicola Ibison and the NCI Management Team who have been my guides through the maze of telly and publishing. Great thanks to Celador Productions and International, who took such a risk taking a big green granny on and turning her into a presenter. Roly Keating at BBC2, thank you for putting a 'real' woman on telly. Thanks to the team at Virgin for publishing this book – a dream come true.

The Celador Production team was full of very special, talented, people. I learnt so much from them and they learned a lot about the environment! In particular I would like to thank (in no particular order) Damon Pattison, Stephanie Weatherill, all the directors, Katie Crawford, Andrew Barron, Angela Norris, especially Dan Riley, Donna Ferry (mother of the family), and Sarah Hayes (quiet angel of efficiency). All our Assistant Producers' were lovely, Debbie Dunnett, Stevie Gower – for your love of life, Jonny Durbridge (kindness himself), Liz Bate (guardian of my fringe), Ollie Stroud (my road buddy). Many thanks also to main cameraman Steve Court and Martin Leberl our soundman – what gents.

No Waste Like Home was about the lives of our contributors, who let us run riot through their homes for weeks. Their kindness, patience, generosity, openness and humour cannot be underestimated. I could not have been more fortunate than to have met them. I would like to thank the Coltman-Wests, as well as the Tibbetts; our student household; Caroline, Jeffrey and the kids; the Lamberts; the Bendell-Jones'; the Henshalls; the Roberts; and finally, Phillip and Andrew who became eco-warriors in their own right – going out and converting so many people to the 'green' side!

The series features many excellent environmentally friendly products and services. Many businesses were extraordinarily generous to us and I would like to single out Jane from The Green Shop in Stroud for making such efforts on behalf of the series.